T0205373

# Springer Theses

Recognizing Outstanding Ph.D. Research

## Aims and Scope

The series "Springer Theses" brings together a selection of the very best Ph.D. theses from around the world and across the physical sciences. Nominated and endorsed by two recognized specialists, each published volume has been selected for its scientific excellence and the high impact of its contents for the pertinent field of research. For greater accessibility to non-specialists, the published versions include an extended introduction, as well as a foreword by the student's supervisor explaining the special relevance of the work for the field. As a whole, the series will provide a valuable resource both for newcomers to the research fields described, and for other scientists seeking detailed background information on special questions. Finally, it provides an accredited documentation of the valuable contributions made by today's younger generation of scientists.

### Theses may be nominated for publication in this series by heads of department at internationally leading universities or institutes and should fulfill all of the following criteria

- They must be written in good English.
- The topic should fall within the confines of Chemistry, Physics, Earth Sciences, Engineering and related interdisciplinary fields such as Materials, Nanoscience, Chemical Engineering, Complex Systems and Biophysics.
- The work reported in the thesis must represent a significant scientific advance.
- If the thesis includes previously published material, permission to reproduce this must be gained from the respective copyright holder (a maximum 30% of the thesis should be a verbatim reproduction from the author's previous publications).
- They must have been examined and passed during the 12 months prior to nomination.
- Each thesis should include a foreword by the supervisor outlining the significance of its content.
- The theses should have a clearly defined structure including an introduction accessible to new PhD students and scientists not expert in the relevant field.

Indexed by zbMATH.

More information about this series at https://link.springer.com/bookseries/8790

Heejae Kim

# Glide-Symmetric $Z_2$ Magnetic Topological Crystalline Insulators

Doctoral Thesis accepted by
Tokyo Institute of Technology, Tokyo, Japan

*Author*
Dr. Heejae Kim
Department of Physics
Tokyo Institute of Technology
Tokyo, Japan

*Supervisor*
Prof. Shuichi Murakami
Department of Physics
Tokyo Institute of Technology
Tokyo, Japan

ISSN 2190-5053 ISSN 2190-5061 (electronic)
Springer Theses
ISBN 978-981-16-9079-2 ISBN 978-981-16-9077-8 (eBook)
https://doi.org/10.1007/978-981-16-9077-8



This Springer imprint is published by the registered company Springer Nature Singapore Pte Ltd.
The registered company address is: 152 Beach Road, #21-01/04 Gateway East, Singapore 189721, Singapore

# Supervisor's Foreword

Symmetry is an important concept in describing complex behaviours in condensed materials, because its effects survive various complications in real materials. Recent enormous progress in the field of topological materials has been showing us intriguing and nontrivial interplay between topology and symmetry. Crystalline systems such as electronic systems and metamaterials allow rich varieties of topological phases, depending on their symmetries. Among various symmetries in solids, nonsymmorphic symmetries such as glide and screw symmetries are nontrivial and particularly interesting, because the fractional translation in these symmetries leads to an intertwined nature of wavefunctions over the Brillouin zone.

This book is concerned with theoretical investigations of fundamental properties of topological phases in crystals protected by glide symmetry. This book, based on the dissertation of Dr. Heejae Kim, theoretically investigates fundamental properties of these phases and applies the results to a design of topological photonic crystals. It is shown that by adding other spatial symmetries, the formula of the topological invariant for the glide symmetry becomes drastically simplified, and is expressed as a form of symmetry-based indicator. This property fascillitates search and design of topological phases in electronic systems and in metamaterials, as discussed in detail in the book. The present theory is applicable to various systems with a wide class of space groups, and can be a powerful tool for understanding and designing topological materials and metamaterials.

Tokyo, Japan
December 2021

Prof. Shuichi Murakami

**Parts of this thesis have been published in the following journal articles:**

1. Heejae Kim, Hengbin Cheng, Ling Lu, and Shuichi Murakami
   *Theoretical analysis of glide-$Z_2$ magnetic topological photonic crystals*,
   Optics Express **29**(20), 31164–31178 (2021).
   
2. Heejae Kim and Shuichi Murakami
   *Glide-symmetric topological crystalline insulator phase in a nonprimitive lattice*,
   Physical Review B **102**, 195202 (2020).
   
3. Heejae Kim, Ken Shiozaki, and Shuichi Murakami
   *Glide-symmetric magnetic topological crystalline insulators with inversion symmetry*,
   Physical Review B **100**, 165202 (2019).
   
4. Heejae Kim and Shuichi Murakami
   *Emergent spinless Weyl semimetals between the topological crystalline insulator and normal insulator phases with glide symmetry*,
   Physical Review B **93**, 195138 (2016).
   

# Acknowledgements

The process inheriting knowledge from past people, exploring knowledge they have never stepped into, and opening up a new field of knowledge makes a treasure trove of humankind nowdays. The Ph.D. program might be one of the driving forces to get where we are today while we have come a long way. The singularity of my Ph.D. program had happened fortunately when I started to work with my advisor, Prof. Shuichi Murakami. The marvelous thing is that no matter when I am working on, with some good ideas to share or struggling with challenging problems, he always provides brilliant insights and comments. His enthusiasm for physics to broaden humankind knowledge is nature and simple, but always stimulated and spurred me on. A wonderful encounter with him must have been a chance for what I can be eventually stepping out in order to do research on my own. I feel deep appreciation for having had him for completion of this thesis.

During my doctoral course, there are many people I have to acknowledge outside Tokyo Institute of Technology. I am really indebted and thankful to Dr. Ken Shiozaki for the collaboration and inspiring discussions, and I have learned various knowledge from him. He had warmly supported me and his insightful suggestions and comments stimulate me enjoying this field of physics. Similarly, I am very grateful to Prof. Haruki Watanabe for invaluable discussions. Discussions with him are always powerful and I have learned lots of knowledge of symmetry-based indicators from him. I would like to thank Prof. Ling Lu for fruitful discussions. I learned topological photonic crystals from him and I can develop my study with our stimulating discussions. I would also like to thank Dr. Motoaki Hirayama. He always gives me beneficial knowledge and I learned material science from him. I also have to appreciate Dr. Ryo Okugawa. Since he was a member of our group, I always have enjoyed invaluable discussions with him, and he inspired me to study physics of topological materials.

There are many people I would like to appreciate in Tokyo Institute of Technology. I would like to thank my fabulous colleagues for fruitful discussions and kind supports in many occasions during my graduate course in the Murakami group, Dr. Takehito Yokoyama, Dr. Akihiro Okamoto, Dr. Masato Hamada, Dr. Tiantian Zhang, Taiki Yoda, Kazuki Yokomizo, Ryo Takahashi, Ryo Furuta, Tomohiro Inoue, Yusuke

Aihara, Daisuke Hara, Manabu Takeichi, Ken Ohsumi, Yutaro Tanaka, Nobuhiro Arai, Hisayoshi Komiyama, and Hiroaki Yamada. I also deeply thank administrative staffs, Keiko Yamada, and research and educational assistant Rie Tsukui. I wish to thank all members in condensed matter theory group at Tokyo Institute of Technology for great assistance.

I would also like to acknowledge financial support from the Japan Society for the Promotion of Science and CREST, Japan Science and Technology Agency.

Finally, I would like to express my gratitude to my parents, Deokheon Kim and Jihyun Lee, and my sister Sooyun Kim for always kind support and encouragement.

# Contents

# Chapter 1
# Introduction

## 1.1 Background

Over the decades, important roles of the relations between topology and symmetry in modern condensed matter physics have been recognized by the tour de force works. The importance of topology in condensed matter physics is recognized by the discovery of quantum Hall effect [1]. Energy levels in the quantum Hall systems form quantized Landau levels and the systems insulating in bulk, but they host gapless conducting states localized at edges under a strong magnetic field in two-dimensional systems. Since these robust gapless edge states are protected by topology, unless the topological invariant known as the Chern number characterizing the quantum Hall states changes, we cannot adiabatically deform them to conventional quantum states for electrons in solids. Namely, topological invariants which are integrals of wavefunctions encode topology of a given system, and two systems with a different value of topological invariants cannot be continuously deformed to each other. The trigger to encourage researches on topological phases is the discovery of a $Z_2$ topological insulator [2, 3]. The $Z_2$ topological insulators which appear in the presence of time-reversal symmetry have attracted tremendous attention because of their unique and robust surface properties, and they are characterized by the $Z_2$ topological invariant, where anti-unitariness of time-reversal symmetry plays a key role. Realistic materials for the $Z_2$ topological insulator are shortly established by experiments in two-dimensional [4] and three-dimensional [5] systems. Physicists have focused on the relations between topology and symmetry, and they have classified various topological phases ensured by combinations of internal symmetries, such as time-reversal symmetry, particle-hole symmetry, and chiral symmetry [6–8].

Beyond internal symmetries, crystal symmetries of crystalline materials, such as spatial inversion, reflection, and rotational symmetries, play a fundamental role to understand such topological phases. An introduction of crystal symmetries has enriched our knowledge for topological invariants [9, 10]. For instance, the $Z_2$ topological invariant for the $Z_2$ topological insulator has a complicated expression when inversion symmetry is absent, while it is expressed as a simple formula as a product



H. Kim, *Glide-Symmetric $Z_2$ Magnetic Topological Crystalline Insulators*,
Springer Theses, https://doi.org/10.1007/978-981-16-9077-8_1

of parity eigenvalues at time-reversal invariant momenta when inversion symmetry is preserved [9]. In addition, since the seminal proposal of a topological crystalline insulator [11], whose topological phase is ensured by crystal symmetries, various combinations of internal symmetry and crystal symmetries have turned out to give an immense list of new topological phases [12]. In particular, the materials of the SnTe class are revealed to be mirror-symmetric topological crystalline insulators, and they are the first material realization for the topological crystalline insulators protected by crystal symmetries [13–16]. Besides symmorphic symmetries, topological phases with nonsymmorphic symmetries have been vigorously discussed. Due to the nonsymmorphic nature, they host exotic surface states, such as Möbius or hourglass type [17–20]. Furthermore, a new class of topological crystalline insulators, dubbed higher-order topological insulators, hosting gapless boundary states emerging in less than $d - 1$ dimension for a $d$-dimensional bulk insulating system, has also been revealed.

The concept of topology was extended from gapped (insulating) states to gapless (metal or semimetal) states, so-called topological semimetal phases hosting gapless points or line nodes at generic momentum ensured by topological reasons. Studies on topological semimetal phases not only enriches our understanding for bulk topology but led us to new concepts of several exotic particles, such as Weyl (Dirac) fermions in a Weyl (Dirac) semimetal phase. Even though most of topological gapless points appear at a generic momentum, symmetry helps us realizing materials for topological semimetal phases [21]. Moreover, topological nodal-line semimetals are closely associated with symmetries; mirror or glide symmetry, or inversion and time-reversal symmetries without spin-orbit couplings.

Superconductivity is also an ideal field to study topological phases since the band structures of superconductors have a gap or gapless nodes naturally, and some topological superconducting phases are predicted to exhibit Majorana fermions [7]. Furthermore, there is no reason to restrict ourselves to electronic systems. A photonic crystal is one of the best playgrounds to investigate physics of topological materials [22, 23]. Because of exotic properties of bosonic systems, they have much potential for realizations and applications. In the similar context, we can also expand our arguments to magnons, phonons, and so on.

For comprehensive classifications of these topological phases, an approach based on the $K$-theory, which is a mathematical approach for gapped systems, has been adopted [24]. By narrowing down our interest to space-group symmetry which crystalline systems belong to, a general formalism to determine whether a given crystallographic system is topologically trivial or nontrivial only from information of symmetry have been attracting research interest, such as topological quantum chemistry [25, 26] and symmetry-based indicators [27, 28]; in these approaches, topology of the bands and their compatibility relations are studied in the context of topological phases protected by space-group symmetries. Topological quantum chemistry gives minimal band connections from representation theory which an atomic insulator satisfies, and it has led us to a catalogue having all the topologically different connections. Symmetry-based indicators are particularly useful in diagnosing topologically distinct band structures from the combinations of irreducible representations at high-symmetry momenta. These approaches focus on different aspects of topological phases. In the $K$-theory approach, one can comprehensively classify nontrivial phases, whereas an explicit formula for a topological invariant does not follow immediately from the theory. On the other hand, the symmetry-based indicator can reveal only the topological phases characterized by combinations of irreducible representations. Thus, this theory cannot capture topological phases which cannot be known only from the irreducible representations. Thus even with these powerful tools, one cannot reach a full understanding of nontrivial topological phases protected by space-group symmetry, and there is much room for further investigation.

In the present thesis, we focus on the glide-$Z_2$ topological crystalline insulator without time-reversal symmetry [17, 18]. A glide operation is a product of a reflection and a fractional translational operations, and it is nonsymmorphic. Such topological phases are characterized by the $Z_2$ topological invariant [17, 18] and the Chern number associated with the normal vector of the glide plane. The nature of a $Z_2$ topology for the magnetic glide-symmetric systems is very different from that for the time-reversal-symmetric $Z_2$ topological insulators, ensured by the antiunitary property of the time-reversal operation, because of an absence of antiunitary symmetry in magnetic glide-symmetric systems. The reason we have been interested in this topological phase is as follows. First, glide symmetry yields $Z_2$ topological phases in systems with no internal symmetry whereas other symmetries can have (mirror) Chern insulating phases characterized by an integer topological invariant [29]. Second, even though the first topological phenomenon in condensed matter physics was discovered in magnetic systems, namely the quantum Hall effect, there is still few proposals for magnetic topological materials because it is challenging to break time-reversal symmetry in electronic systems without significant spin-orbit coupling or by strong magnetic fields. Therefore, the more we understand this $Z_2$ topological phase with glide symmetry, the more we can deepen our knowledge for topological phases in magnetic systems and nonsymmorphic symmetry. Since the glide symmetry is contained in many space groups, this glide-symmetric topological crystalline insulator phase, ensured by the $Z_2$ topological invariant, can exist in many space groups as well.

To reach our goal, we first study phase transitions between the glide-$Z_2$ magnetic topological crystalline insulator and the normal insulator phases in spinless systems when glide symmetry is preserved in the phase transition. In general cases in such a phase transition, we find that a spinless Weyl semimetal phase should appear between two distinct topological phases, and the behavior of Weyl nodes is associated with the change of the glide-$Z_2$ topological invariant. We construct a general theory describing such a phase transition based on an effective model, and then support our scenario by performing numerical calculations. Next, we study topological phases when inversion symmetry is introduced in magnetic systems with glide symmetry, to see an intriguing interplay between topology and glide symmetry. If we add inversion symmetry, we show that the $Z_2$ topological invariant characterizing glide-symmetric systems can be solely expressed in terms of the irreducible representations at high-symmetry points in momentum space. In particular, we find that the glide-symmetric $Z_2$ magnetic topological crystalline insulator is directly related to a higher-order topological insulator ensured by inversion symmetry. We also construct all invariants which characterize topology of layer constructions for the systems with glide and inversion symmetries, in a similar way as in Ref. [30]. This construction is solely based on real-space geometries of layers. We then show that these invariants completely agree with the set of topological invariants constructed from the topology in momentum space. Finally, we give a strategy for manipulation of the glide-$Z_2$ topological phase based on the relationship between space-group representations and band structures, and discuss its consequences in photonic crystals. We find that the irreducible representations at the particular high-symmetry point determine whether a given photonic crystal is topological or not, and by putting dielectrics at some of Wyckoff positions we can realize the band structure with a set of corresponding irreducible representations.

## 1.2 Outline of Thesis

The organization in this thesis is the following. In Chap. 2, we warm up by reviewing the physics of Berry phase, connection, and curvature, band theory, symmetry and representations theory, and several topological phases of matter. Basically we focus on how the band structure can be affected by topology and symmetry, what makes topological phases different from trivial phases, how we can distinguish them, and what properties emerge. Chapter 3 is devoted to a theory for phase transitions that the Weyl semimetal phase emerges between the topologically trivial and nontrivial phases in spinless glide-symmetric magnetic systems. In Chap. 4, we study the fate of the glide-$Z_2$ invariant when inversion symmetry is added. We find that with this additional inversion symmetry, topological invariants are solely expressed from the irreducible representations at high-symmetry points in momentum space, both for a primitive lattice and a nonprimitive lattice. In Chap. 5, we construct real-space topological invariants by a layer construction in magnetic systems to support our results in Chap. 4, and we also construct simple tight-binding models exhibiting the glide-$Z_2$ topological phase and the Chern insulator phase. In Chap. 6, we argue topological photonic crystals ensured by glide symmetry and we propose a way to realize such a photonic crystal in the context of Wyckoff positions. We summarize our conclusions of the previous chapters in Chap. 7.

## References

1. Klitzing Kv, Dorda G, Pepper M (1980) Phys Rev Lett 45:494. https://doi.org/10.1103/PhysRevLett.45.494
2. Hasan MZ, Kane CL (2010) Rev Mod Phys 82:3045. https://doi.org/10.1103/RevModPhys.82.3045
3. Qi XL, Zhang SC (2011) Rev Mod Phys 83:1057. https://doi.org/10.1103/RevModPhys.83.1057

4. König M, Wiedmann S, Brüne C, Roth A, Buhmann H, Molenkamp LW, Qi XL, Zhang SC (2007) Science 318(5851):766. https://doi.org/10.1126/science.1148047
5. Hsieh D, Qian D, Wray L, Xia Y, Hor YS, Cava RJ, Hasan MZ (2008) Nature 452(7190):970
6. Schnyder AP, Ryu S, Furusaki A, Ludwig AWW (2008) Phys Rev B 78:195125. https://doi.org/10.1103/PhysRevB.78.195125
7. Kitaev A (2009) In: AIP conference proceedings, vol 1134, pp 22–30
8. Ryu S, Schnyder AP, Furusaki A, Ludwig AW (2010) New J Phys 12(6):065010
9. Fu L, Kane CL (2007) Phys Rev B 76:045302. https://doi.org/10.1103/PhysRevB.76.045302
10. Teo JCY, Fu L, Kane CL (2008) Phys Rev B 78:045426. https://doi.org/10.1103/PhysRevB.78.045426
11. Fu L (2011) Phys Rev Lett 106:106802. https://doi.org/10.1103/PhysRevLett.106.106802
12. Chiu CK, Teo JCY, Schnyder AP, Ryu S (2016) Rev Mod Phys 88:035005. https://doi.org/10.1103/RevModPhys.88.035005
13. Hsieh TH, Lin H, Liu J, Duan W, Bansil A, Fu L (2012) Nat Commun 3:982. https://doi.org/10.1038/ncomms1969
14. Tanaka Y, Ren Z, Sato T, Nakayama K, Souma S, Takahashi T, Segawa K, Ando Y (2012) Nat Phys 8:800. https://doi.org/10.1038/nphys2442
15. ...Xu SY, Liu C, Alidoust N, Neupane M, Qian D, Belopolski I, Denlinger J, Wang Y, Lin H, Wray L, Landolt G, Slomski B, Dil J, Marcinkova A, Morosan E, Gibson Q, Sankar R, Chou F, Cava R, Bansil A, Hasan M (2012) Nat Commun 3:1192. https://doi.org/10.1038/ncomms2191
16. Dziawa P, Kowalski BJ, Dybko K, Buczko R, Szczerbakow A, Szot M, Lusakowska E, Balasubramanian T, Wojek BM, Berntsen MH, Tjernberg O, Story T (2012) Nat Mater 11:1023. https://doi.org/10.1038/nmat3449
17. Fang C, Fu L (2015) Phys Rev B 91:161105(R). https://doi.org/10.1103/PhysRevB.91.161105
18. Shiozaki K, Sato M, Gomi K (2015) Phys Rev B 91:155120. https://doi.org/10.1103/PhysRevB.91.155120
19. Wang Z, Alexandradinata A, Cava RJ, Bernevig BA (2016) Nature 532(7598):189
20. Ma J, Yi C, Lv B, Wang Z, Nie S, Wang L, Kong L, Huang Y, Richard P, Zhang P, Yaji K, Kuroda K, Shin S, Weng H, Bernevig BA, Shi Y, Qian T, Ding H (2017) Sci Adv 3(5). https://doi.org/10.1126/sciadv.1602415
21. Murakami S, Hirayama M, Okugawa R, Miyake T (2017) Sci Adv 3(5):e1602680
22. Lu L, Joannopoulos JD, Soljačić M (2014) Nat Photonics 8(11):821
23. Ozawa T, Price HM, Amo A, Goldman N, Hafezi M, Lu L, Rechtsman MC, Schuster D, Simon J, Zilberberg O, Carusotto I (2019) Rev Mod Phys 91:015006. https://doi.org/10.1103/RevModPhys.91.015006
24. Freed DS, Moore GW (2013) In: Annales Henri Poincaré, vol 14. Springer, pp 1927–2023
25. Bradlyn B, Elcoro L, Cano J, Vergniory M, Wang Z, Felser C, Aroyo M, Bernevig BA (2017) Nature 547(7663):298
26. Kruthoff J, de Boer J, van Wezel J, Kane CL, Slager RJ (2017) Phys Rev X 7:041069. https://doi.org/10.1103/PhysRevX.7.041069
27. Po HC, Vishwanath A, Watanabe H (2017) Nat Commun 8(1):50
28. Watanabe H, Po HC, Vishwanath A (2018) Sci Adv 4(8):eaat8685
29. Shiozaki K, Sato M, Gomi K (2018) arXiv:1802.06694
30. Song Z, Zhang T, Fang Z, Fang C (2018) Nat Commun 9(1):3530

# Chapter 2
# Topology, Symmetry, and Band Theory of Materials

We review basic properties of topology and symmetry in band theory and the brief history of topological phases of matter in single-particle systems in the present chapter. First, we explain general properties of Berry phase, Berry connection, and Berry curvature and how these quantities are encoded in band theory described by Bloch electrons. We also explain how they correspond to the Chern number, as an example of topological invariants, associated with Wannier functions localized in real space. We also introduce Bloch-like states for photonic crystals. Second, we argue how symmetry affects these topological quantities. In particular, based on space-group symmetry that crystals belong to, we show an interplay between symmetry and band topology focusing on the symmetry-based indicators among the methods describing band topology. Third, we give several topological phases in single-particle physics which are characterized by topological invariants and we demonstrate their exotic topological surface states in a broad context of topological matters beyond electronic systems such as topological photonics.

## 2.1 Band Theory and Topology

We start with giving a review on physics of Berry phase, Berry connection, and Berry curvature and band theory in this section.



H. Kim, *Glide-Symmetric $Z_2$ Magnetic Topological Crystalline Insulators*, Springer Theses, https://doi.org/10.1007/978-981-16-9077-8_2

### 2.1.1 *Physics of Berry Phase, Berry Connection, and Berry Curvature*

A Berry phase is a geometric phase which naturally appears whenever the Hamiltonian depends on an external parameter [1]. We devote this section to explain basic properties of the Berry phase. Consider a Hamiltonian dependent on certain parameters $\lambda = (\lambda_1, \lambda_2, \dots) \in \mathbb{R}_n$. Namely, we write

$$H = H(\lambda), \tag{2.1}$$

and the Hamiltonian $H(\lambda)$ governs the Schrödinger equation,

$$H(\lambda)|\phi(\lambda)\rangle = E(\lambda)|\phi(\lambda)\rangle, \tag{2.2}$$

where $E$ is an eigenenergy and $|\phi(\lambda)\rangle$ is an eigenstate. For simplicity, there is no degeneracy between eigenstates $|\phi(\lambda)\rangle$.

Let us consider the phase difference between the eigenstates at $\lambda_1$ and $\lambda_2$ when the states adiabatically evolve from $\lambda_1$ to $\lambda_2$, and we define this quantity by $-\mathrm{Im}\log\langle\phi(\lambda_1)|\phi(\lambda_2)\rangle$. Now, we suppose sets of parameters $\lambda_n = \lambda_0, \lambda_1, \dots, \lambda_N$ making a closed path, i.e., $\lambda_0 = \lambda_N$. Then, the phase difference of the ground states along this closed path is given by

$$\gamma = -\mathrm{Im}\log\left[\langle\phi(\lambda_0)|\phi(\lambda_1)\rangle\langle\phi(\lambda_1)|\phi(\lambda_2)\rangle\cdots\langle\phi(\lambda_{N-1})|\phi(\lambda_{N=0})\rangle\right], \tag{2.3}$$

known as the *discrete* Berry phase. We can easily extend this idea to a *continuous* expression for the Berry phase by taking the continuum limit in a closed path of the parameters $\lambda$ as

$$\gamma = \oint d\lambda \mathcal{A}(\lambda), \tag{2.4}$$

where $\mathcal{A}(\lambda) \equiv \langle\phi(\lambda)|i\partial_\lambda|\phi(\lambda)\rangle$ is the so-called *Berry connection*. The Berry phase is gauge-invariant whereas the Berry connection is *not*. Let us see this by taking another wavefunction with an arbitrary real phase $\chi(\lambda)$ via a gauge transformation,

$$|\phi'(\lambda)\rangle = e^{i\chi(\lambda)}|\phi(\lambda)\rangle, \tag{2.5}$$

and we find

$$\mathcal{A}'(\lambda) = \langle\phi'(\lambda)|i\partial_\lambda|\phi'(\lambda)\rangle = \langle\phi(\lambda)|i\partial_\lambda|\phi(\lambda)\rangle - \frac{d\chi(\lambda)}{d\lambda} = \mathcal{A}(\lambda) - \frac{d\chi(\lambda)}{d\lambda}. \tag{2.6}$$

This implies the Berry connection depends on the gauge. On the other hand, since $|\phi'(\lambda_0)\rangle = |\phi'(\lambda_N)\rangle$ for a closed path $\lambda_0 = \lambda_N$, the phase factor must be $\chi(\lambda_N) = \chi(\lambda_0) + 2\pi n$ ($n \in \mathbb{Z}$), and it leads to

$$\int_{\lambda_0}^{\lambda_N} d\lambda \frac{d\chi(\lambda)}{d\lambda} = \chi(\lambda_N) - \chi(\lambda_0) = 2\pi n. \tag{2.7}$$

Therefore, this immediately yields $\gamma' = \gamma + 2\pi n$ which implies that the Berry phase $\gamma$ is gauge-invariant modulo $2\pi$.

One can also define an open-path Berry phase, the phase difference between two wavefunctions at $\lambda_i$ and $\lambda_f$ in continuum systems, as

$$\gamma = \int_{\lambda_i}^{\lambda_f} d\lambda \mathcal{A}(\lambda), \tag{2.8}$$

where $\lambda$ is a scalar parameter running from $\lambda_i$ to $\lambda_f$. Obviously, an open-path Berry phase (2.8) is *not* gauge-invariant, whereas Eq. (2.4) is, because the Berry connection depends on the gauge. By introducing the gauge $\chi(\lambda)$ in Eq. (2.5), the Berry phase $\gamma$ changes by $\chi(\lambda_f) - \chi(\lambda_i)$.

Apart from the formulation of the Berry phase, here we explain how this geometric Berry phase relates to adiabatic dynamics encoded by the time-dependent Schrödinger equation. The eigenstates of the Hamiltonian $H(\lambda)$ for a given $\lambda$ are given by

$$H(\lambda)|\phi_n(\lambda)\rangle = E_n|\phi_n(\lambda)\rangle. \tag{2.9}$$

Now we consider how the eigenstate $|\phi_n(\lambda)\rangle$ evolves when the parameter $\lambda$ varies in time evolution. At time $t = 0$, we set the initial state $|\Phi(t = 0)\rangle = |\phi_n(\lambda(t = 0))\rangle$. For the time-dependent Schrödinger equation when the system evolves adiabatically given by

$$i\hbar \frac{\partial}{\partial t}|\Phi(t)\rangle = H(\lambda)|\Phi(t)\rangle, \tag{2.10}$$

this initial state evolves as

$$|\Phi(t)\rangle = e^{i\theta(t)}|\phi_n(\lambda(t))\rangle. \tag{2.11}$$

From Eqs. (2.10) and (2.11), we obtain

$$i\hbar e^{i\theta(t)} \left( i\frac{d\theta(t)}{dt} + \frac{\partial}{\partial t} \right) |\phi_n(\lambda(t))\rangle = E_n(\lambda(t))e^{i\theta(t)}|\phi_n(\lambda(t))\rangle, \tag{2.12}$$

and by acting $\langle\phi_n(\lambda(t))|$ from the left, we get

$$\frac{d\theta(t)}{dt} = i\langle\phi_n(\lambda(t))|\partial_t|\phi_n(\lambda(t))\rangle - \frac{1}{\hbar}E_n(\lambda(t)). \tag{2.13}$$

Hence, the phase factor is

$$\theta(t) = -\frac{1}{\hbar}\int_0^t dt' E_n(\lambda(t')) + i\int_0^t dt' \langle\phi_n(\lambda(t'))|\partial_{t'}|\phi_n(\lambda(t'))\rangle. \tag{2.14}$$

This implies that the state at $t$ is written as

$$|\Phi(t)\rangle = e^{i\gamma(t)}e^{-i\int_0^t dt' E_n(\lambda(t'))/\hbar}|\phi_n(\lambda(t))\rangle, \tag{2.15}$$

$$\gamma(t) = i\int_0^t dt' \langle\phi_n(\lambda(t'))|\partial_{t'}|\phi_n(\lambda(t'))\rangle. \tag{2.16}$$

The factor $e^{-i\int_0^t dt' E_n(\lambda(t'))/\hbar}$ in Eq. (2.15) is the dynamical phase expected for solutions of the time-dependent Schrödinger equation. Meanwhile, the Berry phase $\gamma(t)$ is geometric phase which arises from the adiabatic ansatz that the parameter $\lambda$ varies adiabatically with time $t$. We note that the Berry phase $\gamma(t)$ always takes a real value from the orthonormalization of $\phi_n(\lambda)$. The Berry phase is not gauge-invariant in general, but for a periodic case with $\lambda(t=0) = \lambda(t=T)$, we get

$$\gamma(T) = i\int_0^T dt' \frac{\partial\lambda}{\partial t}\langle\phi_n(\lambda(t'))|\partial_\lambda|\phi_n(\lambda(t'))\rangle = \oint d\lambda \mathcal{A}_n(\lambda). \tag{2.17}$$

Therefore, it becomes gauge invariant. It does not depend on time-dependence of the parameters $\lambda(t)$, but it only depends on the trajectory of $\lambda(t)$ in the $\lambda$-space.

From the similarity between the Berry connection and the gauge potential in electromagnetism, it is useful to define a *Berry curvature* as an anti-symmetric second-rank tensor

$$\mathcal{F}_{n,\mu\nu}(\lambda) = \partial_{\lambda_\mu}\mathcal{A}_{n,\nu}(\lambda) - \partial_{\lambda_\nu}\mathcal{A}_{n,\mu}(\lambda) = -2\mathrm{Im}\langle\partial_{\lambda_\mu}\phi_n(\lambda)|\partial_{\lambda_\nu}\phi_n(\lambda)\rangle. \tag{2.18}$$

We note that the components of the Berry curvature are gauge-invariant with respect to the gauge transformation (2.5). Then, it is possible to express the Berry phase $\gamma$ in terms of the Berry curvature by using the Stokes' theorem

$$\gamma(C) = \oint_C d\boldsymbol{\lambda}\cdot\mathcal{A} = \int_S ds_\mu \wedge ds_\nu \mathcal{F}_{\mu\nu}, \tag{2.19}$$

where $ds_\mu \wedge ds_\nu$ is an area element on the surface $S$ in the parameter space encompassed by the path $C$. Therefore, the Berry phase precisely corresponds to the flux of the Berry curvature through $S$.

We introduce a useful alternative expression for the Berry curvature. By using the first-order perturbation theory, an expression for the derivative of the $|\phi_n(\lambda)\rangle$ explicitly reads

$$|\partial_{\lambda_\nu}\phi_n(\lambda)\rangle = \sum_{m\neq n} \frac{\langle\phi_m(\lambda)|\partial_{\lambda_\mu}H|\phi_n(\lambda)\rangle}{E_n - E_m}|\phi_m(\lambda)\rangle, \tag{2.20}$$

and by plugging in Eq. (2.18), we obtain

$$\mathcal{F}_{n,\mu\nu}(\lambda) = -2\mathrm{Im}\sum_{m\neq n}\frac{\langle\phi_n(\lambda)|\partial_{\lambda_\mu}H|\phi_m(\lambda)\rangle\langle\phi_m(\lambda)|\partial_{\lambda_\nu}H|\phi_n(\lambda)\rangle}{(E_n - E_m)^2}. \tag{2.21}$$

This expression is a gauge invariant form, and with this formula it is easily to understand how the energy differences between the $n$th band and other bands affect the Berry curvature.

Up to this point, we assume the systems without degeneracy. One can also define the *non-Abelian* Berry phase, connection, and curvature arising in multiband systems. Assume there is $N$-fold degeneracy of an eigenvalue $E_n$ at any point $\lambda$. The most general form of the wavefunction for Eq. (2.10) reads

$$|\Phi(t)\rangle = \sum_{\alpha=1}^{N} c_{n\alpha}(t)|\phi_{n\alpha}(\lambda(t))\rangle, \tag{2.22}$$

under the adiabatic assumption once again. From Eqs. (2.10) and (2.22), we obtain

$$c_{\mathbf{n}}(t) = c_{\mathbf{n}}(0)\mathcal{T}\exp\left[\int_0^t dt'\left(-\frac{i}{\hbar}E_n(\lambda(t'))\mathbb{I}_{N\times N} + i\mathcal{A}^N(\lambda(t'))\right)\right] \tag{2.23}$$

where $\mathcal{T}$ is the time-ordering operator, $\mathbb{I}_{N\times N}$ is the $N\times N$ identity matrix, and $\mathcal{A}^N$ is a non-Abelian Berry connection whose components are given by

$$\mathcal{A}^N_{n,\alpha\beta}(\lambda(t)) \equiv \langle\phi_{n\alpha}(\lambda(t))|i\partial_t|\phi_{n\beta}(\lambda(t))\rangle. \tag{2.24}$$

Let us check gauge dependence of the non-Abelian Berry connection. Under the gauge transformation $|\phi'_n(\lambda)\rangle = U^N|\phi_n(\lambda)\rangle$ by a $N\times N$ unitary matrix $U^N$, we find

$$\left(\mathcal{A}^N_n\right)' = \left(U^N\right)^{-1}\mathcal{A}^N_n U^N + i\left(U^N\right)^{-1}dU^N, \tag{2.25}$$

and this is not gauge-covariant. In a periodic case, with the parameter $\lambda(t)$ satisfying $\lambda(0) = \lambda(T)$, the time-evolution unitary operator $\mathcal{U}$ defined by $|\Phi(T)\rangle = \mathcal{U}|\Phi(0)\rangle$ is given by

$$\mathcal{U} = \exp\left[-\frac{i}{\hbar}\int_0^T dt\,E_n(\lambda(t))\right]\mathcal{P}\exp\left[i\oint d\lambda\mathcal{A}^N_n(\lambda)\right]. \tag{2.26}$$

where $\mathcal{P}$ is a path-ordering operator. The first term is the Abelian dynamical factor, and the second term is known as the non-Abelian Berry phase. The non-Abelian Berry phase is generally gauge-dependent. However, its trace (known as *Wilson loop*) is indeed gauge-invariant and behaves similarly to a Berry phase for a single band. The non-Abelian Berry curvature is defined as

$$\mathcal{F}^{N}_{n,\mu\nu}(\lambda) = \partial_{\lambda_\mu}\mathcal{A}^{N}_{n,\nu} - \partial_{\lambda_\nu}\mathcal{A}^{N}_{n,\mu} + i\left[\mathcal{A}^{N}_{n,\mu}, \mathcal{A}^{N}_{n,\nu}\right]. \tag{2.27}$$

The last term is necessary to ensure gauge covariance.

### 2.1.2 Bloch Hamiltonian

So far, we have considered the Berry phase, Berry connection, and Berry curvature defined in a system with a generic parameter $\lambda$. Here we focus on how these properties are encoded in the Bloch wavefunctions in crystals which we are interested in. We start with the Bloch's theorem. The Schrödinger equation for a single-particle wavefunction is

$$H\psi = \left(-\frac{\hbar^2}{2m}\nabla^2 + V(\mathbf{r})\right)\psi = E\psi. \tag{2.28}$$

In a three-dimensional (3D) crystal, because the potential is periodic $V(\mathbf{r}+\mathbf{a}) = V(\mathbf{r})$ for all primitive lattice vectors $\mathbf{a}$ in a Bravais lattice, the system is invariant under any of the translation operators $T_{\mathbf{a}}$ obeying $T_{\mathbf{a}}f(\mathbf{r}) = f(\mathbf{r}-\mathbf{a})$. Thus, the Hamiltonian commutes with a translation operator, $[H, T_{\mathbf{a}}] = 0$. Moreover, the translation operators for all the primitive lattice vectors commute with one another. Since $T_{\mathbf{a}}$ is a unitary operator, its eigenvalue is turned out to $e^{-i\mathbf{k}\cdot\mathbf{a}}$ where $\mathbf{k}$ is a constant, and is called Bloch wavevector. Therefore, an eigenstate $\psi$ of $H$ is also an eigenstate of $T_{\mathbf{a}}$:

$$T_{\mathbf{a}}\psi(\mathbf{r}) = \psi(\mathbf{r}+\mathbf{a}) = e^{i\mathbf{k}\cdot\mathbf{r}}\psi(\mathbf{r}), \tag{2.29}$$

and this is known as Bloch's theorem.

One can define cell-periodic Bloch functions $u_{\mathbf{k}}$ by

$$\psi(\mathbf{r}) = e^{i\mathbf{k}\cdot\mathbf{r}}u_{\mathbf{k}}(\mathbf{r}), \tag{2.30}$$

obeying periodic boundary conditions $u_{\mathbf{k}}(\mathbf{r}+\mathbf{a}) = u_{\mathbf{k}}(\mathbf{r})$. Because $\psi(\mathbf{r})$ is an eigenstate of the Hamiltonian $H$ in Eq. (2.28), by using Eq. (2.30) we obtain the $k$-dependent Bloch Hamiltonian for cell-periodic Bloch functions

$$\mathcal{H}_{\mathbf{k}} = e^{-i\mathbf{k}\cdot\mathbf{r}}He^{i\mathbf{k}\cdot\mathbf{r}} \tag{2.31}$$

such that

$$\mathcal{H}_{\mathbf{k}}|u_{\mathbf{k}}(\mathbf{r})\rangle = E_{\mathbf{k}}|u_{\mathbf{k}}(\mathbf{r})\rangle. \tag{2.32}$$

For the Hamiltonian in Eq. (2.28) as a simplest example, it becomes

$$\mathcal{H}_{\mathbf{k}} = \frac{(-i\hbar\nabla + \hbar\mathbf{k})^2}{2m} + V(\mathbf{r}). \tag{2.33}$$

Any single-particle operator $O$ can be also converted into a $k$-dependent operator, $O_{\mathbf{k}} = e^{-i\mathbf{k}\cdot\mathbf{r}} O e^{i\mathbf{k}\cdot\mathbf{r}}$, which yields

$$\langle O \rangle = \frac{1}{N} \sum_{n\mathbf{k}} \langle \psi_{n\mathbf{k}} | O | \psi_{n\mathbf{k}} \rangle_{\text{cell}} = \frac{1}{N} \sum_{n\mathbf{k}} \langle u_{n\mathbf{k}} | O_{\mathbf{k}} | u_{n\mathbf{k}} \rangle, \tag{2.34}$$

where the suffix *cell* indicates that the expectation values are calculated in an unit cell.

Let us take a parameter for the Berry phase as the Bloch vector $\lambda = \mathbf{k}$. Then, the Berry phase for the Bloch states are

$$\gamma = \oint d\mathbf{k} \cdot \mathcal{A}_n(\mathbf{k}), \tag{2.35}$$

and the corresponding Berry connections and curvatures for the Bloch states can be written as

$$\mathcal{A}_{n,i}(\mathbf{k}) = i \langle u_{n\mathbf{k}} | \partial_{k_i} | u_{n\mathbf{k}} \rangle, \tag{2.36}$$
$$\mathcal{F}_{n,ij}(\mathbf{k}) = \partial_{k_i} \mathcal{A}_{n,j} - \partial_{k_j} \mathcal{A}_{n,i}, \tag{2.37}$$

where $i$, $j$ run through a Cartesian index and $|u_{n\mathbf{k}}\rangle$ is the $n$th Bloch state. Note that we have adopted cell-periodic Bloch functions $|u_{n\mathbf{k}}\rangle$ instead of the Bloch functions $|\psi_{n\mathbf{k}}\rangle$ to define these quantities. Since all of the cell-periodic Bloch functions $|u_{n\mathbf{k}}\rangle$ satisfy the same simple periodic condition $u_{\mathbf{k}}(\mathbf{r}) = u_{\mathbf{k}}(\mathbf{r} + \mathbf{a})$, inner products between vectors at different $\mathbf{k}$, or derivatives with respect to $\mathbf{k}$, are well defined in the Brillouin zone (BZ). It is useful to choose $|u_{n\mathbf{k}}\rangle$ to argue the band topology which is generally given by the map $\mathcal{H}(\mathbf{k})$ from the BZ. The Bloch states $|u_{n\mathbf{k}}\rangle$ have a $U(1)$ gauge freedom: $|u'_{n\mathbf{k}}\rangle = e^{i\theta(\mathbf{k})}|u_{n\mathbf{k}}\rangle$. Therefore, the Berry connection depends on the gauge choice,

$$\mathcal{A}'_{n,i}(\mathbf{k}) = \mathcal{A}_{n,i}(\mathbf{k}) - \partial_{k_i}\theta(\mathbf{k}). \tag{2.38}$$

Meanwhile, the Berry phase is gauge-invariant modulo $2\pi$ and the Berry curvature is fully gauge-invariant, which directly follows from the expression in Eq. (2.21)

$$\mathcal{F}_{n,ij} = -2\text{Im} \sum_{m \neq n} \frac{\langle u_{n\mathbf{k}} | \partial_{k_i} \mathcal{H}_{\mathbf{k}} | u_{m\mathbf{k}} \rangle \langle u_{m\mathbf{k}} | \partial_{k_j} \mathcal{H}_{\mathbf{k}} | u_{n\mathbf{k}} \rangle}{(E_{n\mathbf{k}} - E_{m\mathbf{k}})^2}. \tag{2.39}$$

Here we briefly mention several properties of the Berry curvature in $k$-space. If the crystal has time-reversal symmetry, the Berry curvature is odd, $\mathcal{F}_n(\mathbf{k}) = -\mathcal{F}_n(-\mathbf{k})$, while it is even in the presence of inversion symmetry, $\mathcal{F}_n(\mathbf{k}) = \mathcal{F}_n(-\mathbf{k})$. Thus, the Berry curvature identically vanishes in the presence of both of time-reversal symmetry and inversion symmetry. The Berry curvature is also transformed by a given spatial symmetry of the crystal, and we will discuss this in detail in the following section. We can also define a *monopole density* as

$$\rho_n(\mathbf{k}) = \nabla_{\mathbf{k}} \cdot \mathcal{F}_n(\mathbf{k}). \tag{2.40}$$

This quantity is always a linear combination of delta functions with their coefficients quantized in the unit of $2\pi$. Their singularities correspond to the Weyl fermion in the BZ.

In a multiband system, the non-Abelian Berry connections and curvatures are

$$\mathcal{A}_{mn,i}(\mathbf{k}) = i\langle u_{m\mathbf{k}}|\partial_{k_i}|u_{n\mathbf{k}}\rangle, \tag{2.41}$$

$$\mathcal{F}_{mn,ij}(\mathbf{k}) = \partial_{k_i}\mathcal{A}_{mn,j} - \partial_{k_j}\mathcal{A}_{mn,i} + i\left[\mathcal{A}_{mn,i}, \mathcal{A}_{mn,j}\right], \tag{2.42}$$

where $\left[\mathcal{A}_{mn,i}, \mathcal{A}_{mn,j}\right] = \sum_s \left(\mathcal{A}_{ms,i}\mathcal{A}_{sn,j} - \mathcal{A}_{ms,j}\mathcal{A}_{sn,i}\right)$. Let us check their gauge-dependence. For a multiband $\mathrm{U}(N)$ gauge transformation,

$$|u'_{n\mathbf{k}}\rangle = \sum_m U_{mn}(\mathbf{k})|u_{m\mathbf{k}}\rangle, \tag{2.43}$$

the non-Abelian Berry connection is transformed by

$$\mathcal{A}'_{mn,i}(\mathbf{k}) = \left(U^\dagger \mathcal{A}_i U\right)_{mn} + i\left(U^\dagger \partial_{k_i} U\right)_{mn}, \tag{2.44}$$

and this is not gauge-covariant. On the other hand, by plugging in Eq. (2.42), we obtain $\mathcal{F}'_{mn,ij} = \left(U^\dagger \mathcal{F}_{ij} U\right)_{mn}$. Thus, the non-Abelian Berry curvature is gauge-covariant.

### 2.1.3 Chern Number

We have studied that the Berry phase is a geometric phase which naturally appears in quantum mechanics when the system is deformed adiabatically, and the one around the closed path $P$ corresponds to the flux of the Berry curvature through the surface having the boundary $P$. In fact, these quantities are deeply associated with the (first) *Chern number*, one of the topological invariants. Apart from the general argument of the Chern theorem, here we discuss the Chern number for the Bloch Hamiltonian.

The manifold of the BZ in 2D can be regarded as a torus. A fully gapped Bloch Hamiltonian defined in the BZ can be characterized by the Chern number

$$C = \frac{1}{2\pi}\int_{\mathrm{BZ}} d^2k\ \mathrm{tr}\mathcal{F}_{xy}(\mathbf{k}). \tag{2.45}$$

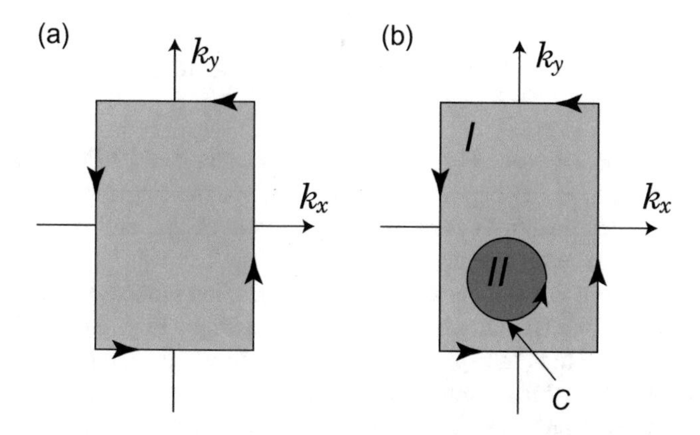


**Fig. 2.1** Regions for the integral of Berry curvature in whole Brillouin zone (BZ). **a** There is no singularity in whole BZ. **b** There is a singularity in BZ for the gauge I. Near the singularity (region II), we take another gauge II and $C$ is the boundary between regions I and II

The integral is taken over the whole region of the BZ in 2D. It is obvious that the Chern number is gauge invariant, and that the Chern number is quantized to be an integer. We show this in the following. By using the Stokes' theorem, the Chern number can be rewritten as

$$C = \frac{1}{2\pi} \oint_{\partial \mathrm{BZ}} d\mathbf{k} \cdot \mathrm{tr}\mathcal{A}(\mathbf{k}), \tag{2.46}$$

where $\partial$BZ is the boundary of the BZ. Thanks to the periodicity of the BZ, this integral vanishes (Fig. 2.1a). However, if there is a singularity in BZ, one cannot determine a single gauge in the entire BZ and (2.46) becomes invalid. Namely, if we employ a certain gauge, i.e., a certain choice of the phase of the wavefunction, the phase becomes inconsistent around certain $\mathbf{k}$ point, where the phase winds around. Such point is an singularity mentioned above. To calculate the Chern number in such cases, we divide the BZ into two regions: one has no singularity (I) and the other is enclosing the singularity (II) shown in Fig. 2.1b. In the region II, we employ a different choice of gauge, such that it has no singularity in the region II. We set the wavefunctions in region I and II as $|u^{\mathrm{I}}_{n\mathbf{k}}\rangle$ and $|u^{\mathrm{II}}_{n\mathbf{k}}\rangle$, respectively, and corresponding Berry connections are denoted by $\mathcal{A}^{\mathrm{I}}(\mathbf{k})$ and $\mathcal{A}^{\mathrm{II}}(\mathbf{k})$. Then, from the Stokes' theorem, the Chern number is calculated as a contour integral on loop $C$ which is the boundary between two regions I and II (Fig. 2.1b),

$$C = \frac{1}{2\pi} \oint_{C} d\mathbf{k} \cdot \left(\mathrm{tr}\mathcal{A}^{\mathrm{II}}(\mathbf{k}) - \mathrm{tr}\mathcal{A}^{\mathrm{I}}(\mathbf{k})\right). \tag{2.47}$$

Because the wavefunctions $|u^{\mathrm{I}}_{n\mathbf{k}}\rangle$ and $|u^{\mathrm{II}}_{n\mathbf{k}}\rangle$ are related by the gauge transformation on the loop $C$ as $|u^{\mathrm{II}}_{n\mathbf{k}}\rangle = e^{i\theta_n(\mathbf{k})}|u^{\mathrm{I}}_{n\mathbf{k}}\rangle$, we obtain $\mathcal{A}^{\mathrm{II}}(\mathbf{k}) = \mathcal{A}^{\mathrm{I}}(\mathbf{k}) - \partial_{\mathbf{k}}\mathrm{diag}(\theta_1(\mathbf{k}), \ldots, \theta_n(\mathbf{k}))$ and

$$C = \frac{1}{2\pi} \oint_{\mathrm{C}} \sum_{n=1}^{N} d\mathbf{k} \cdot \frac{\partial \theta_n(\mathbf{k})}{\partial \mathbf{k}} = \frac{1}{2\pi} \sum_{n=1}^{N} [\theta_n(\mathbf{k})]_{\mathrm{C}}, \tag{2.48}$$

in which $N$ is the total number of bands we consider and $[\theta_n(\mathbf{k})]_{\mathrm{C}}$ denotes the change of the phase $\theta_n(\mathbf{k})$ on the loop C. Because of the single-valuedness of the wavefunctions, this quantity has an integer multiple of $2\pi$, and hence the Chern number is shown to be an integer.

In fact, the Chern number corresponds to the winding number representing how many times the Bloch Hamiltonian winds around a singularity in the BZ. Suppose the Bloch Hamiltonian is fully gapped in the BZ. Because the BZ is a torus, being a closed manifold, the Chern number on this BZ torus can be defined. Let us consider a two-band model represented by a Hamiltonian $H = \mathbf{d}(\mathbf{k}) \cdot \boldsymbol{\sigma}$, where $\boldsymbol{\sigma}$ is a Pauli matrix. Then for the lower band, we have

$$\mathcal{A}_\mu(\mathbf{k}) = \left(\partial_{k_\mu} d_\alpha(\mathbf{k})\right) \mathcal{A}_\alpha(\mathbf{d}), \tag{2.49}$$

$$\mathcal{F}_{\mu\nu}(\mathbf{k}) = -\frac{1}{2} \hat{\mathbf{d}} \cdot \left(\partial_\mu \hat{\mathbf{d}} \times \partial_\nu \hat{\mathbf{d}}\right), \tag{2.50}$$

and

$$C = -\frac{1}{4\pi} \int_{S^2} d^2k \, \hat{\mathbf{d}} \cdot \left(\partial_\mu \hat{\mathbf{d}} \times \partial_\nu \hat{\mathbf{d}}\right). \tag{2.51}$$

This is a powerful result since it gives a geometric interpretation of the Chern number as the wrapping number of a closed sphere. Recall $\hat{\mathbf{d}}(\mathbf{k})$ defines a wrapping of a unit sphere, and the Chern number is calculated as the total solid angle subtended by $\mathbf{d}(\mathbf{k})$ divided by $4\pi$. It leads us to an intuitive understanding that the Chern number is regarded as a winding number. If there is a singularity, i.e., a nonzero monopole density (or a Weyl fermion), inside the BZ torus, the Chern number is nonzero, and it is zero otherwise.

### 2.1.4 Hybrid Wannier Centers

When we compute the Berry phase or Berry connection, it is technically useful to employ Wannier functions. At first, we consider an isolated band and its Bloch functions $|\psi_{n\mathbf{k}}\rangle$ with a smooth and periodic gauge. By a Fourier transform, the Wannier function localized at a site $\mathbf{R}$ is defined by

$$|w_{n\mathbf{R}}\rangle = \frac{V}{(2\pi)^d} \int_{\mathrm{BZ}} d\mathbf{k} e^{-i\mathbf{k}\cdot\mathbf{R}} |\psi_{n\mathbf{k}}\rangle, \tag{2.52}$$

and its inverse transform is

$$|\psi_{n\mathbf{k}}\rangle = \sum_{\mathbf{R}} e^{i\mathbf{k}\cdot\mathbf{R}}|w_{n\mathbf{R}}\rangle, \tag{2.53}$$

where $V$ is the unit cell volume. One can easily see that the Wannier functions form an orthonormal and complete set due to Eq. (2.30), and two Wannier functions are associated by the translation operator; $|w_{n\mathbf{R}}\rangle = T_{\mathbf{R}}|w_{n\mathbf{0}}\rangle$. The Hilbert spaces spanned by the Wannier functions and the Bloch functions are equal. Namely, the projector onto the band $n$ is

$$P_n = \frac{V}{(2\pi)^d}\int_{\mathrm{BZ}} d\mathbf{k}|\psi_{n\mathbf{k}}\rangle\langle\psi_{n\mathbf{k}}| = \sum_{\mathbf{R}}|w_{n\mathbf{R}}\rangle\langle w_{n\mathbf{R}}|. \tag{2.54}$$

The Wannier functions have a gauge degree of freedom, a freedom of gauge choice, because the phases for the Bloch functions are not determined uniquely.

Let us consider the center of the Wannier functions, by taking the expectation value of the position operator in the home unit cell $\mathbf{R} = \mathbf{0}$,

$$\bar{\mathbf{r}}_n = \langle w_{n\mathbf{0}}|\mathbf{r}|w_{n\mathbf{0}}\rangle. \tag{2.55}$$

By using the definition of the Wannier functions, it becomes

$$\bar{\mathbf{r}}_n = \frac{V}{(2\pi)^d}\int_{\mathrm{BZ}} d\mathbf{k}\langle u_{n\mathbf{k}}|i\partial_{\mathbf{k}}|u_{n\mathbf{k}}\rangle = \frac{V}{(2\pi)^d}\int_{\mathrm{BZ}} d\mathbf{k}\mathcal{A}_n(\mathbf{k}). \tag{2.56}$$

In particular, in 1D systems with the lattice constant $a$, we have

$$\bar{x}_n = \frac{a}{2\pi}\int_{-\pi/a}^{\pi/a} dk\langle u_{nk}|i\partial_k|u_{nk}\rangle = \frac{a}{2\pi}\gamma_n, \tag{2.57}$$

in which $\gamma_n$ is the Berry phase for the $n$th band. Namely, the Berry phase for the $n$th band in $k$-space precisely corresponds to a Wannier center in real space.

In 1D systems, the Wannier functions are maximally localized in real space and those are eigenfunctions of the projected position operator, which means that they are gauge independent, while they are not in 2D or 3D systems since the projected position operators do not commute each other. Nonetheless, we can compute the Berry phases in higher dimensions from the *hybrid* Wannier functions that are Wannier-like in one of the dimensions and Bloch-like in remaining dimensions [2]. Suppose a 3D system and choose the $z$ direction for constructing the Wannier function only, i.e.,

$$|w_{nl_z}\rangle = \frac{c}{2\pi}\int dk_z e^{-ik_z l_z c}|\psi_{n\mathbf{k}}\rangle, \tag{2.58}$$

where $l_z$ is a site index and $c$ is a lattice constant along the $z$ direction. Then, the hybrid Wannier centers in the home unit cell is

$$\bar{z}_n(k_x, k_y) = \langle w_{n0}|z|w_{n0}\rangle = \frac{c}{2\pi}\int_{-\pi/c}^{\pi/c} dk_z \mathcal{A}_n(\mathbf{k}). \tag{2.59}$$

This is exactly the Berry phase along the $k_z$ direction on the $k_x$-$k_y$ plane.

Technically, the hybrid Wannier centers are very useful in calculating the Chern number numerically. By choosing a smooth and periodic gauge along $k_x$ direction, and wrapping the BZ in the $k_y$ direction, but not in $k_x$,[1] Eq. (2.45) can be transformed as

$$\begin{aligned} C &= \frac{1}{2\pi}\int_{\mathrm{BZ}} dk_x dk_y \mathrm{tr}\mathcal{F}_{xy}(\mathbf{k}) = \frac{1}{2\pi}\left[\oint dk_y \mathrm{tr}\mathcal{A}_y(k_x = 2\pi) - \oint dk_y \mathrm{tr}\mathcal{A}_y(k_x = 0)\right] \\ &= \frac{1}{b}\sum_n \left[\bar{y}_n(2\pi) - \bar{y}_n(0)\right], \end{aligned} \tag{2.60}$$

in which $b$ is a lattice constant along the $y$ direction. Namely, by tracking the Berry phase in the $k_y$ direction, we can obtain the Chern number on a torus.

In fact, the individual hybrid Wannier center corresponds to the eigenvalues of the Wilson loop $W(k) = \mathrm{tr}\left(\mathcal{P}\exp[i\oint dk\mathcal{A}_n(k)]\right)$ as we stated in Sect. 2.1.1 if the gauge is chosen such that these hybrid Wannier functions are maximally localized, i.e.,

$$\bar{x}_n = \frac{a}{2\pi}\arg(\lambda_n), \tag{2.61}$$

up to possible reordering [3]. $\lambda_n$ denotes an eigenvalue of the Wilson loop.

### 2.1.5 Photonic Crystal

One can derive Bloch-like states from the Maxwell equations instead of the Schrödinger equation. This occurs in *photonic crystal* systems when the light propagates in a medium with the periodic dielectric function $\varepsilon(\mathbf{r}+\mathbf{a}) = \varepsilon(\mathbf{r})$ [4]. Recall the Maxwell equations

$$\nabla\cdot\mathbf{D} = \rho, \tag{2.62}$$

$$\nabla\cdot\mathbf{B} = 0, \tag{2.63}$$

$$\nabla\times\mathbf{E} + \frac{\partial\mathbf{B}}{\partial t} = 0, \tag{2.64}$$

$$\nabla\times\mathbf{H} - \frac{\partial\mathbf{D}}{\partial t} = \mathbf{j}, \tag{2.65}$$

[1] Namely, it is regarded as a cylinder geometry different from a torus or a sphere we have dealt with.

where **E** and **H** are the electric and magnetic fields, **D** and **B** are the electric displacement and magnetic flux density, and $\rho$ and **j** are the free charge and free current densities, respectively. Here we assume that the photonic crystal is a dielectric medium with the magnetic permeability to be that of vacuum. Namely, $\mathbf{D}(\mathbf{r}) = \varepsilon_0\varepsilon(\mathbf{r})\mathbf{E}(\mathbf{r})$ and $\mathbf{B} = \mu_0\mathbf{H}$, where $\varepsilon(\mathbf{r}+\mathbf{a}) = \varepsilon(\mathbf{r})$ for all the primitive lattice vectors **a**. We also assume $\rho = 0$ and $\mathbf{j} = 0$. Under these conditions, the Maxwell equations then yield

$$\nabla \cdot [\varepsilon(\mathbf{r})\mathbf{E}(\mathbf{r}, t)] = 0, \tag{2.66}$$

$$\nabla \cdot \mathbf{H}(\mathbf{r}, t) = 0, \tag{2.67}$$

$$\nabla \times \mathbf{E}(\mathbf{r}, t) + \mu_0 \frac{\partial \mathbf{H}(\mathbf{r}, t)}{\partial t} = 0, \tag{2.68}$$

$$\nabla \times \mathbf{H}(\mathbf{r}, t) - \varepsilon_0\varepsilon(\mathbf{r}) \frac{\partial \mathbf{E}(\mathbf{r}, t)}{\partial t} = 0. \tag{2.69}$$

By eliminating either **E** or **H** in Eqs. (2.68) and (2.69), we obtain

$$\frac{1}{\varepsilon(\mathbf{r})} \nabla \times [\nabla \times \mathbf{E}(\mathbf{r}, t)] + \varepsilon_0\mu_0 \frac{\partial^2 \mathbf{E}(\mathbf{r}, t)}{\partial t^2} = 0, \tag{2.70}$$

$$\nabla \times \left[\frac{1}{\varepsilon(\mathbf{r})} \nabla \times \mathbf{H}(\mathbf{r}, t)\right] + \varepsilon_0\mu_0 \frac{\partial^2 \mathbf{H}(\mathbf{r}, t)}{\partial t^2} = 0. \tag{2.71}$$

For an electromagnetic wave with frequency $\omega$,

$$\mathbf{E}(\mathbf{r}, t) = \mathbf{E}(\mathbf{r})e^{-i\omega t}, \tag{2.72}$$

$$\mathbf{H}(\mathbf{r}, t) = \mathbf{H}(\mathbf{r})e^{-i\omega t}, \tag{2.73}$$

we have eigenvalue equations for $\mathbf{E}(\mathbf{r})$ and $\mathbf{H}(\mathbf{r})$

$$\hat{\mathcal{L}}^E \mathbf{E}(\mathbf{r}) \equiv \frac{1}{\varepsilon(\mathbf{r})} \nabla \times (\nabla \times \mathbf{E}(\mathbf{r})) = \frac{\omega^2}{c^2} \mathbf{E}(\mathbf{r}), \tag{2.74}$$

$$\hat{\mathcal{L}}^H \mathbf{H}(\mathbf{r}) \equiv \nabla \times \left(\frac{1}{\varepsilon(\mathbf{r})} \nabla \times \mathbf{H}(\mathbf{r})\right) = \frac{\omega^2}{c^2} \mathbf{H}(\mathbf{r}), \tag{2.75}$$

where $\omega$ is a eigenfrequency and $c = 1/\sqrt{\varepsilon_0\mu_0}$ is the speed of light in vacuum. Note that $\hat{\mathcal{L}}^H$ is a Hermitian operator whereas $\hat{\mathcal{L}}^E$ is not. Below, we hence focus on the eigenvalue problem for $\mathbf{H}(\mathbf{r})$ for mathematical convenience. Nonetheless, those are associated with each other by recovering $\mathbf{E}(\mathbf{r})$ or $\mathbf{H}(\mathbf{r})$ in Eqs. (2.68) and (2.69).

Now we expand $\mathbf{H}(\mathbf{r})$ to the Bloch-like form because $\varepsilon(\mathbf{r})$ is spatially periodic. Suppose a 3D periodic system with three primitive lattice vectors $\mathbf{a}_{i=1,2,3}$ and corresponding reciprocal lattice vectors $\mathbf{b}_{i=1,2,3}$. Then, the modes labeled by the Bloch wave vector **k** can be defined as

$$\mathbf{H}_{\mathbf{k}}(\mathbf{r}) = e^{i\mathbf{k}\cdot\mathbf{r}}\mathbf{u}_{\mathbf{k}}(\mathbf{r}), \tag{2.76}$$

where $\mathbf{u_k}(\mathbf{r} + \mathbf{a}_i) = \mathbf{u_k}(\mathbf{r})$ is a periodic function as a similar manner in Eq. (2.30). Plugging in $\mathbf{H_k}$ for Eq. (2.75), we obtain

$$\hat{\mathcal{L}}^H_{\mathbf{k}} \mathbf{u_k}(\mathbf{r}) \equiv (i\mathbf{k} + \nabla) \times \frac{1}{\varepsilon(\mathbf{r})} (i\mathbf{k} + \nabla) \times \mathbf{u_k}(\mathbf{r}) = \frac{\omega^2_{\mathbf{k}}}{c^2} \mathbf{u_k}(\mathbf{r}). \tag{2.77}$$

This eigenvalue equation is precisely the same form of Eq. (2.32) for periodic Bloch wave vectors $\mathbf{k}$. Thus, it yields band structure of the photonic crystal and one may argue topology of the photonic bands. Note that the Bloch functions $\mathbf{u_k}(\mathbf{r})$ are vector functions whereas those for Bloch electrons are scalar. Obviously, these Bloch functions should be transverse waves since one of Maxwell equations, Eq. (2.66) or (2.67), yields $(i\mathbf{k} + \nabla) \cdot \mathbf{u_k} = 0$ and forbids longitudinal waves.

## 2.2 Symmetry Analysis and Notation

The band topology is closely associated with symmetry. The crystals have corresponding space-group symmetry and internal symmetry, such as time-reversal, particle-hole or chiral symmetries. We here briefly manifest the constraint on the band structures under those symmetries and review several important properties exploited in the following chapters.

### 2.2.1 *Internal Symmetry: Ten-Fold Classification*

When we solve the Schrödinger equation, by using unitary transformations, one may make the Hamiltonian block-diagonal, and eventually it is decomposed into irreducible blocks. Internal symmetries, such as time-reversal symmetry, particle-hole symmetry, and chiral symmetry, forming ten-fold Altland-Zirnbauer symmetry classes [5], impose certain constraints on an irreducible Hamiltonian.

Time-reversal symmetry $\Theta$, particle-hole symmetry $\Xi$, and chiral symmetry $\Pi$ act on $\psi_i$ as

$$\Theta \psi_i \Theta^{-1} = [U_\Theta]_{ij}\, \psi_j, \quad \Theta i \Theta^{-1} = -i, \tag{2.78}$$

$$\Xi \psi_i \Xi^{-1} = [U_\Xi]^*_{ij}\, \psi^\dagger_j, \quad \Xi i \Xi^{-1} = i, \tag{2.79}$$

$$\Pi \psi_i \Pi^{-1} = [U_\Xi U_\Theta]_{ij}\, \psi^\dagger_j, \tag{2.80}$$

in which $U_\Theta$ and $U_\Xi$ are unitary matrices $U^\dagger_\Theta U_\Theta = U_\Theta U^\dagger_\Theta = 1$ and $U^\dagger_\Xi U_\Xi = U_\Xi U^\dagger_\Xi = 1$, respectively, and we have used the relation $\Pi = \Theta\Xi$. When a non-interacting Hamiltonian $\hat{H} = \sum \psi^\dagger_i H_{ij} \psi_j$ is invariant under such symmetries, we have

**Table 2.1** Altland-Zirnbauer symmetry classes. The numbers $\pm 1$ refers to the value of $\Theta^2$ or $\Xi^2$ in each class

| AZ | A | AIII | AI | BDI | D | DIII | AII | CII | C | CI |
|---|---|---|---|---|---|---|---|---|---|---|
| $\Theta$ | | | $+1$ | $+1$ | | $-1$ | $-1$ | $-1$ | | $+1$ |
| $\Xi$ | | | | $+1$ | $+1$ | $+1$ | | $-1$ | $-1$ | $-1$ |
| $\Pi$ | | 1 | | 1 | | 1 | | 1 | | 1 |
| $d=1$ | 0 | $\mathbb{Z}$ | 0 | $\mathbb{Z}$ | $\mathbb{Z}_2$ | $\mathbb{Z}_2$ | 0 | $2\mathbb{Z}$ | 0 | 0 |
| $d=2$ | $\mathbb{Z}$ | 0 | 0 | 0 | $\mathbb{Z}$ | $\mathbb{Z}_2$ | $\mathbb{Z}_2$ | 0 | $2\mathbb{Z}$ | 0 |
| $d=3$ | 0 | $\mathbb{Z}$ | 0 | 0 | 0 | $\mathbb{Z}$ | $\mathbb{Z}_2$ | $\mathbb{Z}_2$ | 0 | $2\mathbb{Z}$ |

$$U_\Theta^\dagger H^* U_\Theta = H, \quad U_\Xi^\dagger H^T U_\Xi = -H, \quad \left(U_\Xi^* U_\Theta^*\right)^\dagger H \left(U_\Xi^* U_\Theta^*\right) = -H, \tag{2.81}$$

respectively. In the Bloch basis in Eq. (2.30), the Bloch Hamiltonian $\mathcal{H}_\mathbf{k} \equiv e^{-i\mathbf{k}\cdot\mathbf{r}} H e^{i\mathbf{k}\cdot\mathbf{r}}$ satisfies

$$\Theta\mathcal{H}_\mathbf{k}\Theta^{-1} = \mathcal{H}_{-\mathbf{k}}, \quad \Xi\mathcal{H}_\mathbf{k}\Xi^{-1} = -\mathcal{H}_{-\mathbf{k}}, \quad \Pi\mathcal{H}_\mathbf{k}\Pi^{-1} = -\mathcal{H}_\mathbf{k}. \tag{2.82}$$

These relations imply that the eigenenergies $E_\mathbf{k}$ and $E_{-\mathbf{k}}$ are symmetric under $\Theta$, $E_\mathbf{k}$ and $-E_{-\mathbf{k}}$ under $\Xi$, and $E_\mathbf{k}$ and $-E_\mathbf{k}$ under $\Pi$, respectively, such that $\mathcal{H}_\mathbf{k}|u_\mathbf{k}\rangle = E_\mathbf{k}|u_\mathbf{k}\rangle$. Since $\Theta^2 = \pm 1$ and $\Xi^2 = \pm 1$, those eigenstates may or may not belong the same energy band.

These three fundamental internal symmetries and their combinations classify such a Hamiltonian into ten classes, dubbed ten-fold Altland-Zirnbauer symmetry classes, summarized in Table 2.1. It shows what kind of topological gapped phases are realized in 1D, 2D or 3D systems.

Henceforth, we do not deal with systems with particle-hole symmetry or chiral symmetry. Here we focus on time-reversal symmetry and how it affects the Berry connections and curvatures. From Eq. (2.82), the bands at $\mathbf{k}$ and $-\mathbf{k}$ are related to $|u_{n,-\mathbf{k}}\rangle = U^*_{nm}(\mathbf{k})\Theta|u_{m\mathbf{k}}\rangle$ with an unitary matrix $U(\mathbf{k})$ satisfying $U(-\mathbf{k}) = -U^T(\mathbf{k})$. Then, the Berry connection at $-\mathbf{k}$ is

$$\begin{aligned}
\mathcal{A}_{i,mn}(-\mathbf{k}) &= -\langle u_{m,-\mathbf{k}}|i\partial_{k_i}|u_{n,-\mathbf{k}}\rangle \\
&= -\langle u_{m'\mathbf{k}}|\Theta U_{mm'}(\mathbf{k}) i\partial_{k_i} U^*_{nn'}(\mathbf{k})\Theta|u_{n'\mathbf{k}}\rangle \\
&= U_{mm'}(\mathbf{k})U^\dagger_{n'n}(\mathbf{k})\langle u_{n'\mathbf{k}}|i\partial_{k_i}|u_{m'\mathbf{k}}\rangle - \delta_{m'n'}U_{mm'}(\mathbf{k})i\partial_{k_i}U^*_{nn'}(\mathbf{k}) \\
&= U_{mm'}(\mathbf{k})\mathcal{A}_{i,m'n'}(\mathbf{k})U^\dagger_{n'n}(\mathbf{k}) - U_{ml}(\mathbf{k})i\partial_{k_i}U^\dagger_{ln}(\mathbf{k}),
\end{aligned} \tag{2.83}$$

and its $U(1)$ gauge part is obtained as

$$\mathrm{tr}\mathcal{A}_i(-\mathbf{k}) = \mathrm{tr}\mathcal{A}_i(\mathbf{k}) + i\,\mathrm{tr}\left(U^\dagger(\mathbf{k})\partial_{k_i}U(\mathbf{k})\right). \tag{2.84}$$

It leads the Berry curvature under $\Theta$ by using Eq. (2.42),

$$\mathrm{tr}\mathcal{F}_{ij}(\mathbf{k}) = -\mathrm{tr}\mathcal{F}_{ij}(-\mathbf{k}), \tag{2.85}$$

that is exactly the relation we mentioned in the previous section. Identically, the Chern number vanishes in the 2D BZ with time-reversal symmetry or a 2D time-reversal-symmetric slice of a BZ in higher dimensions. Namely, the Chern number is zero in the presence of $\Theta$, while it can be nonzero in the absence of $\Theta$.

### 2.2.2 Space-Group Symmetry

Apart from these internal symmetries, crystalline materials conserve a set of symmetry operations in crystals known as *space-group symmetry* acting on Hamiltonians locally. Here we briefly introduce basic concepts of space-group symmetry and representation theory.

A Bravais lattice is an infinite periodic array of points constructed from a set of translation vectors. Such a lattice is preserved by some symmetry operations, i.e., point-group symmetry, such as inversion symmetry, rotational symmetry or mirror symmetry. A crystal structure of crystalline materials in three dimensions determines a space group given by a combination of a point-group and translation operators stemming from a Bravais lattice. Usually, to express an element $g$ of space group $\mathcal{G}$, i.e., $g \in \mathcal{G}$, we adopt the *Seitz notation* $g = \{p|\mathbf{a}\}$ where $p$ is a point-group operation, followed by a translation vector $\mathbf{a}$. This acts on $g\mathbf{r} = \{p|\mathbf{a}\}\mathbf{r} = p\mathbf{r} + \mathbf{a}$, and there is a simple but essential rule for combining operations:

$$\{p|\mathbf{a}_p\}\{p'|\mathbf{a}_{p'}\} = \{pp'|\mathbf{a}_p + p\mathbf{a}_{p'}\}. \tag{2.86}$$

Following relations are then immediately obtained:

$$\{p|\mathbf{a}\}^{-1} = \{p^{-1}| - p^{-1}\mathbf{a}\}, \tag{2.87}$$

$$\{p|\mathbf{a}\} = \{E|\mathbf{a}\}\{p|\mathbf{0}\} = \{p|\mathbf{0}\}\{E|p^{-1}\mathbf{a}\}, \tag{2.88}$$

where $E$ is the identity operator. We can write a space group $\mathcal{G}$ in terms of a finite number of left coset representatives of the lattice translation subgroup $T$:

$$\mathcal{G} = \{p_1|\mathbf{a}_{p_1}\}T + \{p_2|\mathbf{a}_{p_2}\}T + \cdots \{p_n|\mathbf{a}_{p_n}\}T, \tag{2.89}$$

in which $n$ is the macroscopic order of $\mathcal{G}$. In general, $\{p|\mathbf{a}_p\}\{p'|\mathbf{a}_{p'}\} = \{pp'|\mathbf{a}_p + p\mathbf{a}_{p'}\}$ is not equivalent to $\{pp'|\mathbf{a}_{pp'}\}$, and $\{p|\mathbf{a}_p\}$ does not compose a group structure. However, sometimes we can choose an appropriate coset representatives as $\mathbf{a}_p + p\mathbf{a}_{p'} = \mathbf{a}_{pp'}$ for all point-group operations and it leads $\{p|\mathbf{a}_p\}\{p'|\mathbf{a}_{p'}\} = \{pp'|\mathbf{a}_{pp'}\}$. In this case, a space group is *symmorphic*. On the other hand, a space group is *nonsymmorphic* if $\mathbf{a}_p + p\mathbf{a}_{p'} - \mathbf{a}_{pp'} \neq 0$. If a space group is a symmorphic group, i.e., all $\mathbf{a}_{p_i}$'s are zero, the wavefunctions have a same periodicity of the BZ; otherwise, some $\mathbf{a}_p$'s are nonzero, and the nonsymmorphic part yields an integer multiple of reciprocal lattice vectors for the periodicity of the wavefunctions.

### 2.2.3 Little Group of k (k-Group) and Its Representation

Let us consider the representation theory of a space group. Let $\mathbf{a}_1$, $\mathbf{a}_2$, and $\mathbf{a}_3$ denote three basis vectors and $\mathbf{b}_1$, $\mathbf{b}_2$, and $\mathbf{b}_3$ denote the corresponding reciprocal vectors such that $\mathbf{a}_i \cdot \mathbf{b}_j = 2\pi\delta_{ij}$ $(i, j = 1, 2, 3)$. Because $T$ consists of elements $\{E|n_1\mathbf{a}_1 + n_2\mathbf{a}_2 + n_3\mathbf{a}_3\}$ where $n_i$ is an integer that satisfies $0 \leq n_i < N_i, i = 1, 2$ or 3 and $N_1$, $N_2$, and $N_3$ are the total numbers of lattice sites along $\mathbf{a}_1$, $\mathbf{a}_2$, and $\mathbf{a}_3$, respectively, imposing $\{E|\mathbf{a}_1\}^{N_1} = \{E|\mathbf{a}_2\}^{N_2} = \{E|\mathbf{a}_3\}^{N_3} = \{E|\mathbf{0}\}$. Then its representations are given by

$$D(\{E|n_1\mathbf{a}_1 + n_2\mathbf{a}_2 + n_3\mathbf{a}_3\}) = e^{-i\mathbf{k}\cdot(n_1\mathbf{a}_1+n_2\mathbf{a}_2+n_3\mathbf{a}_3)}. \tag{2.90}$$

This is a 1D irreducible representation and therefore the representation matrix, its character, and its eigenvalue are the same. Thus, we have the basis function for $D(\{E|\mathbf{a}\})$ as the plane waves $\exp(i\mathbf{k} \cdot \mathbf{r})$ because

$$\{E|\mathbf{a}\} \exp(i\mathbf{k} \cdot \mathbf{r}) = \exp(-i\mathbf{k} \cdot \mathbf{a}) \exp(i\mathbf{k} \cdot \mathbf{r}) = D(\{E|\mathbf{a}\}) \exp(i\mathbf{k} \cdot \mathbf{r}). \tag{2.91}$$

If one uses any function of the form $\psi_{\mathbf{k}}(\mathbf{r}) = e^{i\mathbf{k}\cdot\mathbf{r}} u_{\mathbf{k}}(\mathbf{r})$ where $u_{\mathbf{k}}(\mathbf{r}) = u_{\mathbf{k}}(\mathbf{r} + \mathbf{a})$, it is nothing other than the Bloch's theorem in Eqs. (2.29) and (2.30). In other words, under such a periodic potential, the wavefunctions are always of the Bloch type.

Now we explain the little group of $\mathbf{k}$ (also known as $k$-group). When an element $g = \{p|\mathbf{a}_p\}$ of $\mathcal{G}$ operates on the plane waves, we have $\{p|\mathbf{a}_p\} \exp(i\mathbf{k} \cdot \mathbf{r}) = \exp(ip\mathbf{k} \cdot (\mathbf{r} - \mathbf{a}_p))$. Namely, the wavevector $\mathbf{k}$ is transformed to $p\mathbf{k}$. Among symmetry operations of $\mathcal{G}$, some point-group operations leave a momentum $\mathbf{k}$ invariant up to a reciprocal lattice vector $\mathbf{G}$, i.e., $\mathbf{k} = p\mathbf{k} \pmod{\mathbf{G}}$. Thus, the little group of $\mathbf{k}$ can be obtained as a set of elements of the point group of $\mathbf{k}$:

$$\mathcal{G}_{\mathbf{k}} = \{g \in \mathcal{G} | {}^{\exists}\mathbf{G} \text{ s.t. } g\mathbf{k} = \mathbf{k} + \mathbf{G}\}. \tag{2.92}$$

Since an element of $\mathcal{G}_{\mathbf{k}}$ locally acts on $\mathbf{k}$, the wavefunctions yield representations of $\mathcal{G}_{\mathbf{k}}$. Such representations can be factorized into a direct sum of irreducible representations and express symmetry properties of the band structure.

We proceed to the representation theory for the space group. When the little group $\mathcal{G}_{\mathbf{k}}$ is a symmorphic group, we obtain an irreducible representation of the $\mathcal{G}_{\mathbf{k}}$ as *a small representation*

$$D_{\mathbf{k}}(\{p|\mathbf{R}\}) = e^{i\mathbf{k}\cdot\mathbf{R}} D(p), \tag{2.93}$$

where $\{p|\mathbf{R}\} \in \mathcal{G}_{\mathbf{k}}$, $e^{i\mathbf{k}\cdot\mathbf{R}}$ comes from $T$, with a lattice vector $\mathbf{R}$ in a Bravais lattice, and the representation $D(p)$ merely depends on a point group. Thus, its corresponding character is given by

$$\chi_{\mathbf{k}}(\{p|\mathbf{R}\}) = e^{i\mathbf{k}\cdot\mathbf{R}}\chi(p). \tag{2.94}$$

On the other hand, for the wavevectors **k** at the boundary of the BZ in nonsymmorphic space groups, one may consider projective representations of the point group instead of conventional representations because the representation of a nonsymmorphic space group is not equivalent to the conventional representation of a point group, even though the factor group $\mathcal{G}_{\mathbf{k}}/T_{\mathbf{k}}$ is isomorphic to a point group. Let the matrices $D(p)$ and $D(p')$ be projective representations of elements $\{p|\mathbf{R}\}$ and $\{p'|\mathbf{R}'\}$ of a point group. Then they satisfy

$$D(p)D(p') = \omega_{p,p'}D(pp'), \tag{2.95}$$

where $\omega_{p,p'}$ is a $U(1)$ phase factor that forms a factor system associated with projective representations. Usually, the projective representations are classified into trivial and nontrivial classes by their factor systems. If $\omega_{p,p'} = 1$, the projective representation is a conventional representation; otherwise it is in a class of nontrivial projective representations not included in the conventional character tables. The different $U(1)$ phase factor $\omega_{p,p'}$ from identity stems from nonsymmorphic space groups or from an internal degrees of freedom, for example, spin.

Let us investigate the action of the point-group operations on the Bloch functions. Using $\hat{p}$ denoting the action of $g = \{p|\mathbf{0}\}$ on the Bloch basis, we obtain

$$\hat{p}\left(\mathcal{H}_{\mathbf{k}}(\mathbf{r})|u_{\mathbf{k}}(\mathbf{r})\rangle\right) = \mathcal{H}_{\mathbf{k}}(p^{-1}\mathbf{r})|u_{\mathbf{k}}(p^{-1}\mathbf{r})\rangle = \mathcal{H}_{p\mathbf{k}}(\mathbf{r})\hat{p}|u_{\mathbf{k}}(\mathbf{r})\rangle, \tag{2.96}$$

and consequently, $\hat{p}\mathcal{H}_{\mathbf{k}}\hat{p}^{-1} = \mathcal{H}_{p\mathbf{k}}$ and $E_{p\mathbf{k}} = E_{\mathbf{k}}$ for non-degenerate bases $\hat{p}|\psi_{\mathbf{k}}(\mathbf{r})\rangle = |\psi_{p\mathbf{k}}(\mathbf{r})\rangle$. For the little group $\mathcal{G}_{\mathbf{k}}$ including $g$, $\mathcal{H}_{\mathbf{k}}$ commutes with $\hat{p}$. Therefore, we construct a common basis of wavefunctions for $\mathcal{H}_{\mathbf{k}}$ and $\hat{p}$ as

$$\hat{p}|\phi^{\alpha}_{i,\mathbf{k}}\rangle = \sum_{j}|\phi^{\alpha}_{j,\mathbf{k}}\rangle\left[D^{\alpha}_{\mathbf{k}}(g)\right]_{ji}, \tag{2.97}$$

where $D^{\alpha}_{\mathbf{k}}(g)$ is the $\alpha$th irreducible representation of $\mathcal{G}_{\mathbf{k}}$. Obviously, because the trace of $D^{\alpha}_{\mathbf{k}}(g)$ is identical to its character, the fundamental transformation of the wavefunctions can be obtained from character tables.

Our interest is how the Berry connection and curvature are transformed by the point-group operations. When $p\mathbf{k}$ is $\mathbf{k}$ transformed by $p$ that is a point-group operation belonging $\mathcal{G}_{\mathbf{k}}$, by using a non-Abelian gauge transformation in Eq. (2.43) the bands at $p\mathbf{k}$ and $\mathbf{k}$ are associated with

$$|u_{m,p\mathbf{k}}\rangle = \sum_{m} U^{*}_{mn}(\mathbf{k})\hat{p}|u_{n,\mathbf{k}}\rangle, \tag{2.98}$$

where $U_{nm}$ is the sewing matrix $U_{mn}(\mathbf{k}) = \langle u_{m,p\mathbf{k}}|\hat{p}|u_{n,\mathbf{k}}\rangle$ [6]. Then, we have the Berry connection at $p\mathbf{k}$ [6]

$$\begin{aligned}\mathcal{A}_{i,mn}(p\mathbf{k}) &= \langle u_{m,p\mathbf{k}}|p_{ij}i\partial_{k_j}|u_{n,p\mathbf{k}}\rangle \\ &= p_{ij}\sum_{m'n'}\langle u_{m',\mathbf{k}}|\hat{p}^{-1}U_{mm'}(\mathbf{k})i\partial_{k_j}U^*_{nm'}(\mathbf{k})\hat{p}|u_{n',p\mathbf{k}}\rangle \\ &= p_{ij}\left(U(\mathbf{k})\mathcal{A}_j(\mathbf{k})U^\dagger(\mathbf{k})\right)_{mn} + p_{ij}\left(U(\mathbf{k})i\partial_{k_j}U^\dagger(\mathbf{k})\right)_{mn}, \end{aligned} \tag{2.99}$$

where $p_{ij}$ is a matrix component representing $p$. This is nothing other than a non-Abelian gauge transformation as we have shown in the previous section with the exception of the presence of the prefactor $U^p_{ij}$. Thus, we have the relations [6]

$$\begin{aligned}\partial_{k_i}\mathcal{A}_j(p\mathbf{k}) - \partial_{k_j}\mathcal{A}_i(p\mathbf{k}) &= p_{ii'}p_{jj'}\left[\partial_{k_{i'}}\left(U\mathcal{A}_{j'}U^\dagger + iU\partial_{k_{j'}}U^\dagger\right) - \partial_{k_{j'}}\left(U\mathcal{A}_{i'}U^\dagger + iU\partial_{k_{i'}}U^\dagger\right)\right] \\ &= p_{ii'}p_{jj'}\Big[U(\partial_{k_{i'}}\mathcal{A}_{j'} - \partial_{k_{j'}}\mathcal{A}_{i'})U^\dagger + i(\partial_{k_{i'}}U\partial_{k_{j'}}U^\dagger - \partial_{k_{j'}}U\partial_{k_{i'}}U^\dagger) \\ &\quad +(\partial_{k_{i'}}U)\mathcal{A}_{j'}U^\dagger + U\mathcal{A}_{j'}(\partial_{k_{i'}}U) - (\partial_{k_{j'}}U)\mathcal{A}_{i'}U^\dagger - U\mathcal{A}_{i'}(\partial_{k_{j'}}U)\Big],\end{aligned} \tag{2.100}$$

$$\begin{aligned}[\mathcal{A}_i(p\mathbf{k}), \mathcal{A}_j(p\mathbf{k})] &= p_{ii'}p_{jj'}\left[U\mathcal{A}_{i'}U^\dagger + iU\partial_{k_{i'}}U^\dagger, U\mathcal{A}_{j'}U^\dagger + iU\partial_{k_{j'}}U^\dagger\right] \\ &= p_{ii'}p_{jj'}\Big[U[\mathcal{A}_{i'}, \mathcal{A}_{j'}]U^\dagger + (\partial_{k_{i'}}U\partial_{k_{j'}}U^\dagger - \partial_{k_{j'}}U\partial_{k_{i'}}U^\dagger) \\ &\quad +i\left((\partial_{k_{i'}}U)\mathcal{A}_{j'}U^\dagger + U\mathcal{A}_{j'}(\partial_{k_{i'}}U) - (\partial_{k_{j'}}U)\mathcal{A}_{i'}U^\dagger - U\mathcal{A}_{i'}(\partial_{k_{j'}}U)\right)\Big],\end{aligned} \tag{2.101}$$

and these therefore lead to the Berry curvature under $p$

$$\mathcal{F}_{ij}(p\mathbf{k}) = p_{ii'}p_{jj'}U(\mathbf{k})\mathcal{F}_{i'j'}(\mathbf{k})U^\dagger(\mathbf{k}). \tag{2.102}$$

### *2.2.4 Compatibility Relations*

In the following subsections, we mimic the arguments in Refs. [7, 8] to review symmetry-based indicators. To this end, we briefly explain compatibility relations, Wyckoff positions, and site symmetries following Refs. [7, 8].

The irreducible representations fulfill *compatibility relations* for the connections of band structures in momentum space [9]. For elements $f \in \mathcal{G}, g \in \mathcal{G}_\mathbf{k}$, and $g' \in \mathcal{G}_{g\mathbf{k}}$ where $\mathcal{G}$, $\mathcal{G}_\mathbf{k}$, and $\mathcal{G}_{g\mathbf{k}}$ denote a space group and the little groups of $\mathbf{k}$ and $g\mathbf{k}$, respectively, one can have $g' = fgf^{-1}$. Then, using Eq. (2.97) we obtain

$$\begin{aligned}\hat{g}'(\hat{f}|\phi^\alpha_{i,\mathbf{k}}\rangle) = \omega_{g',f}(\widehat{g'f})|\phi^\alpha_{i,\mathbf{k}}\rangle = \omega_{g',f}(\widehat{fg})|\phi^\alpha_{i,\mathbf{k}}\rangle &= \frac{\omega_{g',f}}{\omega_{f,g}}\hat{f}(\hat{g}|\phi^\alpha_{i,\mathbf{k}}\rangle) \\ &= \rho_{g',f}\hat{f}(|\phi^\alpha_{j,\mathbf{k}}\rangle[D^\alpha_\mathbf{k}(g)]_{ji}) \\ &= (\hat{f}|\phi^\alpha_{j,\mathbf{k}}\rangle)\rho_{g',f}[D^\alpha_\mathbf{k}(f^{-1}g'f)]_{ji},\end{aligned} \tag{2.103}$$

where the $\hat{f}$ denotes the action of $f$ and $\rho_{g',f}$ is defined as

$$\rho_{g',f} = \frac{\omega_{g',f}}{\omega_{f,f^{-1}g'f}} = \pm 1, \tag{2.104}$$

associated with a factor system of the projective representations $\omega_{p,p'} = \pm 1$. Therefore, $\hat{f}|\phi^{\alpha}_{i,\mathbf{k}}\rangle$ is a basis of the irreducible representation $D^{\alpha'}_{g\mathbf{k}}(f) \equiv \rho_{g',f} D^{\alpha}_{\mathbf{k}}(f^{-1}g'f)$.

Now we check how compatibility relations appear in band structures. Assume that an irreducible representation $D^{\alpha}_{\mathbf{k}}$ of $\mathcal{G}_{\mathbf{k}}$ emerges $n^{\alpha}_{\mathbf{k}}$ times. When $g \notin \mathcal{G}_{\mathbf{k}}$, even though $\mathbf{k}$ and $g\mathbf{k}$ are distinct momenta because $g$ changes the momentum $\mathbf{k}$, the irreducible representations $D^{\alpha}_{\mathbf{k}}$ and $D^{\alpha'}_{g\mathbf{k}}$ for those momenta are related to each other as we have shown above, and it leads to

$$n^{\alpha'}_{g\mathbf{k}} = n^{\alpha}_{\mathbf{k}}. \tag{2.105}$$

Namely, the representations at symmetry-related momenta are associated with each other [7].

We can also consider the compatibility relations for adjacent $\mathbf{k}$ vectors. When the momentum $\mathbf{k}$ at the high-symmetry point in momentum space belongs the little group $\mathcal{G}_{\mathbf{k}}$, the little group $\mathcal{G}_{\mathbf{k}+\delta\mathbf{k}}$ for the momentum $\mathbf{k} + \delta\mathbf{k}$ adjacent to $\mathbf{k}$ lowers its symmetry from $\mathcal{G}_{\mathbf{k}}$, i.e., $\mathcal{G}_{\mathbf{k}+\delta\mathbf{k}} \leq \mathcal{G}_{\mathbf{k}}$. Namely, the irreducible representations at $\mathbf{k}$ can be written in terms of those at $\mathbf{k} + \delta\mathbf{k}$. By reducing the irreducible representations, we obtain the compatibility relations

$$n^{\alpha}_{\mathbf{k}+\delta\mathbf{k}} = \sum_{\beta} c^{\mathbf{k},\delta\mathbf{k}}_{\alpha\beta} n^{\beta}_{\mathbf{k}}, \tag{2.106}$$

where $c^{\mathbf{k},\delta\mathbf{k}}_{\alpha\beta}$ is a non-negative integer [8].

As a similar manner, one may also find the compatibility relations when we go from a higher-symmetry space group to a lower-symmetry space group. We will encounter this situation in Chaps. 4 and 6.

### 2.2.5 Wyckoff Positions

Most generally, the information of the little group of $\mathbf{k}$ is encoded in real-space *Wyckoff positions* and those site symmetry representations. By following the similar way to define the little group of $\mathbf{k}$, let us define the little group of $\mathbf{r}$ as the subgroup of $\mathcal{G}$ that leaves $\mathbf{r}$ invariant. It is called *site symmetry* group $\mathcal{G}_{\mathbf{r}}$ and given as a set

$$\mathcal{G}_{\mathbf{r}} = \{h \in \mathcal{G} | ^{\exists}\mathbf{R} \text{ s.t. } h(\mathbf{r}) \equiv p_h\mathbf{r} + \mathbf{a}_h = \mathbf{r} + \mathbf{R}\} \tag{2.107}$$

for a symmetry element $h = \{p_h|\mathbf{a}_h\}$ and a lattice translation vector $\mathbf{R}$. Every point in real space is classified by its site symmetry, and if the points belong the same site symmetry, they have the same Wyckoff position. The set of symmetry-equivalent points in a space group is called a crystallographic orbit. Namely, such a crystallographic orbit $\mathbf{r}_\sigma = g_\sigma \mathbf{r}$ $(\sigma = 1, 2, \ldots, n)$, with $g \notin \mathcal{G}_\mathbf{r}$ but $g \in \mathcal{G}$, of a symmetry site $\mathbf{r}$ modulo lattice translations can be classified by Wyckoff positions. Here $n$ is the multiplicity of the Wyckoff positions.

Next, we introduce states $|\phi^\alpha_{\mathbf{r},i,\mathbf{k}}\rangle$ on $\mathbf{r}$, obeying an irreducible representation $D^\alpha_\mathbf{r}$ of $\mathcal{G}_\mathbf{r}$, i.e.,

$$\hat{h}|\phi^\alpha_{\mathbf{r},i,\mathbf{k}}\rangle = \sum_j |\phi^\alpha_{\mathbf{r},j,p_h\mathbf{k}}\rangle [D^\alpha_\mathbf{r}(h)]_{ji}, \tag{2.108}$$

where $\hat{h}$ is the action of $h$ onto the states, and obviously it satisfies

$$t_\mathbf{a}|\phi^\alpha_{\mathbf{r},i,\mathbf{k}}\rangle = |\phi^\alpha_{\mathbf{r},i,\mathbf{k}}\rangle e^{-i\mathbf{k}\cdot\mathbf{a}} \tag{2.109}$$

for an element of translation lattice group $t_\mathbf{a} \in T$. Then, $g_\sigma$ transforms an orbital to one at different sites:

$$|\phi^\alpha_{\mathbf{r}_\sigma,i,\mathbf{k}}\rangle \equiv \hat{g}_\sigma |\phi^\alpha_{\mathbf{r},i,p_\sigma^{-1}\mathbf{k}}\rangle. \tag{2.110}$$

These concepts are deeply associated with the band structures since we can determine an irreducible representation $D^\alpha_\mathbf{r}$ associated with the site symmetry $\mathcal{G}_\mathbf{r}$ of the Wyckoff position $\mathbf{r}$ considered.

### 2.2.6 *Topological Invariant: Symmetry-Based Indicator*

As the end of this section, we devote this subsection to *symmetry-based indicators* [7, 8] that are related to topological invariants classifying whether the systems are topologically trivial or not. Suppose there are band gaps between the connected valence bands and the conduction bands at the high-symmetry points in momentum space for a given space group $\mathcal{G}$. Then, those valence bands at each high-symmetry point $\mathbf{k}$ are characterized a set of irreducible representations $D^\alpha_\mathbf{k}$ of the little group $\mathcal{G}_\mathbf{k}$, and let $n^\alpha_\mathbf{k}$ denote the number of $D^\alpha_\mathbf{k}$ emerging in the set. Since we assume the existence of the band gap at the high-symmetry points, $n^\alpha_\mathbf{k}$ is well-defined regardless of possible gap closing at generic $\mathbf{k}$ points. Therefore, the set of $n^\alpha_\mathbf{k}$ for all inequivalent high-symmetry points yields a set of integers $\mathbf{b} = \{n^\alpha_\mathbf{k}\}$, which are associated with topology of the valence bands. Obviously, the integers $\{n^\alpha_\mathbf{k}\}$ must satisfy the compatibility relations we have seen above. Under these conditions, all valid band structures $\{\text{BS}\}$ are given by a combination of $\mathbf{b}$, and we call its dimension $d_\text{BS}$ [8]. Namely, within the abelian group generated by the irreducible representations at high-symmetry points, the generators of its subgroup satisfying the compatibility relations form a $d_\text{BS}$-dimensional abelian group [8].

Then, the symmetry-based indicator $X_{\mathrm{BS}}$ is given by the quotients of $\{\mathrm{BS}\}$ (or the $K$-group in the context of $K$-theory [10]) by the abelian groups generated by atomic insulators $\{\mathrm{AI}\}$

$$X_{\mathrm{BS}} = \frac{\{\mathrm{BS}\}}{\{\mathrm{AI}\}}. \tag{2.111}$$

Atomic insulator refers to insulators where electrons are strongly captured by atoms, and they are characterized the set of trivial irreducible representations. One can construct the set of irreducible representations for atomic insulators by using Wyckoff positions and those site symmetries. Recall that an element $g \in \mathcal{G}$ can be decomposed as $g = t_{\mathbf{a}} g_\sigma h$ where $t_{\mathbf{a}} \in T$ and $h \in \mathcal{G}_{\mathbf{r}}$. For $gg_\sigma$, it is decomposed as $gg_\sigma = t_{\mathbf{a}} g_{\sigma'} h$ with $\mathbf{a} = g(\mathbf{r}_\sigma) - \mathbf{r}_{\sigma'}$ [8]. Then, we obtain

$$\begin{aligned}
\hat{g}|\phi^\alpha_{\mathbf{r}_\sigma,i,\mathbf{k}}\rangle = \omega_{g,g_\sigma}(g\hat{g}_\sigma)|\phi^\alpha_{\mathbf{r},i,p_\sigma^{-1}\mathbf{k}}\rangle &= \omega_{g,g_\sigma}(\widehat{t_{\mathbf{a}} g_{\sigma'} h})|\phi^\alpha_{\mathbf{r},i,p_\sigma^{-1}\mathbf{k}}\rangle \\
&= \frac{\omega_{g,g_\sigma}}{\omega_{g_{\sigma'},h}} \hat{t}_{\mathbf{a}} \hat{g}_{\sigma'}(\hat{h}|\phi^\alpha_{\mathbf{r},i,p_\sigma^{-1}\mathbf{k}}\rangle) \\
&= \frac{\omega_{g,g_\sigma}}{\omega_{g_{\sigma'},h}} \sum_{i'} \hat{t}_{\mathbf{a}}(\hat{g}_{\sigma'}|\phi^\alpha_{\mathbf{r},i',p_g p_\sigma^{-1}\mathbf{k}}\rangle)[D^\alpha_{\mathbf{r}}(h)]_{i'i} \\
&= \frac{\omega_{g,g_\sigma}}{\omega_{g_{\sigma'},h}} \sum_{i'} (\hat{t}_{\mathbf{a}}|\phi^\alpha_{\mathbf{r}_{\sigma'},i',p_g\mathbf{k}}\rangle)[D^\alpha_{\mathbf{r}}(h)]_{i'i} \\
&= \sum_{\sigma',i',\mathbf{k}'} |\phi^\alpha_{\mathbf{r}_{\sigma'},i',\mathbf{k}'}\rangle[\mathcal{D}^\alpha_{\mathbf{r}}(g)]_{\sigma'i'\mathbf{k}',\sigma i\mathbf{k}},
\end{aligned} \tag{2.112}$$

where $\sigma$ and $\sigma'$ denote the number of Wyckoff positions, and $i$ and $i'$ denote matrix indices of the irreducible representation for the site symmetry [8]. We define

$$[\mathcal{D}^\alpha_{\mathbf{r}}(g)]_{\sigma'i'\mathbf{k}',\sigma i\mathbf{k}} = \delta'_{\mathbf{r}_{\sigma'},g(\mathbf{r}_\sigma)} \delta'_{\mathbf{k}',p_g\mathbf{k}} e^{-i\mathbf{k}\cdot(g(\mathbf{r}_\sigma)-\mathbf{r}_{\sigma'})} \frac{\omega_{g,g_\sigma}}{\omega_{g_{\sigma'},h}} [D^\alpha_{\mathbf{k}}(h^g_{\sigma',\sigma})]_{i'i}, \tag{2.113}$$

$\delta'_{\mathbf{r}_1,\mathbf{r}_2} = 1$ if $\mathbf{r}_1 = \mathbf{r}_2$ up to a lattice vector, and $h^g_{\sigma',\sigma} \equiv g_{\sigma'}^{-1} t_{\mathbf{r}_{\sigma'}-g(\mathbf{r}_\sigma)} g g_\sigma$ [8]. In the choice of a particular $\mathbf{k}$, $\mathcal{D}^\alpha_{\mathbf{r}}$ yields a representation of the atomic insulator by restricting $\mathcal{G}$ to $\mathcal{G}_{\mathbf{k}}$, and its character is

$$\chi^\alpha_{\mathbf{r},\mathbf{k}}(g) \equiv \mathrm{tr}\mathcal{D}^\alpha_{\mathbf{r}} = \sum_{\sigma=1}^{|W_{\mathbf{r}}|} \delta'_{\mathbf{r}_\sigma,g(\mathbf{r}_\sigma)} e^{-i\mathbf{k}\cdot(g(\mathbf{r}_\sigma)-\mathbf{r}_\sigma)} \frac{\omega_{g,g_\sigma}}{\omega_{g_{\sigma'},h}} \chi^\alpha_{\mathbf{r}}(h^g_{\sigma,\sigma}), \tag{2.114}$$

where $W_{\mathbf{r}}$ is the number of Wyckoff positions [8]. This tool is powerful to diagnose whether the system is topological or not only from information of wavefunctions at the high-symmetry points.

## 2.3 Topological Phases of Matter

In this section, we review several topological phases of matter. Topological phases for non-interacting particles are mainly classified into topological insulating phases and topological semimetal phases. Such topological insulating phases are characterized by topological invariants and surface states stemming from bulk-boundary correspondence. We introduce Chern insulators, (time-reversal) $Z_2$ topological insulators, topological crystalline insulators, and higher-order topological insulators. For these examples, we argue relations between topology and symmetry. Then we explain topological semimetals, mainly Weyl semimetals. Finally, we give several topics on topological matters in a broad context beyond electronic systems.

### 2.3.1 *Chern Insulator*

Historically, the quantum Hall effect is the first phenomenon described by topological band theory in condensed matter physics. When electrons are placed in two dimensions under the strong magnetic field, their cyclotron motions form quantized Landau levels with energy $E = \hbar\omega_c(m + 1/2)$ at low temperatures, where $\omega_c$ is the cyclotron frequency and $m$ is a non-negative integer. Consequently, the quantum Hall effect can occur only in a system with broken time-reversal symmetry by the external magnetic fields. The cyclotron motions of electrons cancel each other in bulk, and the Hall current flows only on edges of the system. Instead of the external magnetic field, the magnetization can also induce the quantum Hall effect, dubbed *quantum anomalous Hall effect*, in an insulator without Landau levels. Such insulators that establish quantum anomalous Hall effects are called *Chern insulators* and the seminal proposal of the Chern insulators is given by Haldane in 1988 [11]. Experimentally, the quantum anomalous Hall effect is confirmed in thin films of Cr-doped $(Bi, Sb)_2Te_3$ on a $SrTiO_3$ dielectric substrate in 2013 [12] and in ultracold fermionic atoms in a periodically modulated optical honeycomb lattice mimicking the Haldane model in 2014 [13].

For the quantum Hall states, the Hall conductivity is quantized as

$$\sigma_{xy} = -N\frac{e^2}{h}. \tag{2.115}$$

The band structure of the quantized Landau levels has gaps similar to those of an ordinary insulator. The quantum Hall state, however, belongs to a different topological class from an ordinary insulator, classified as class A in the Altland-Zirnbauer classes (Table 2.1) from the absence of time-reversal symmetry, and the Chern number which is a topological invariant, as we studied in Sect. 2.1.3, characterizing the quantum Hall state. The Hall conductivity is written by using the Kubo fomula [14] as

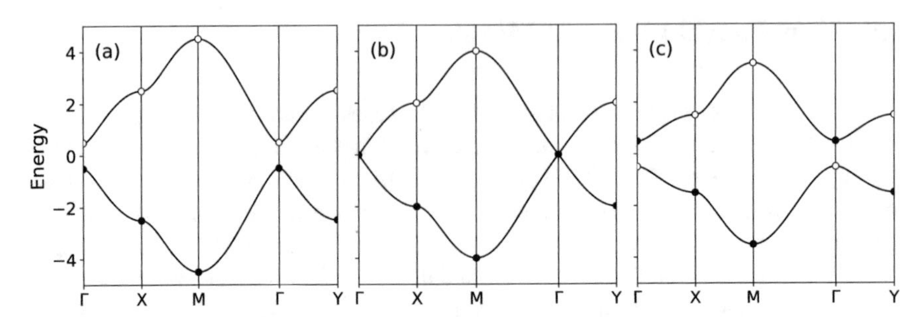


**Fig. 2.2** Bulk band structures for the tight-binding models for the Chern insulator of Eq. (2.118) with **a** $m = -2.5$; **b** $m = -2$; **c** $m = -1.5$. Filled (open) circles mark states of a common character; a band inversion at Γ is evident in **c**

$$\sigma_{xy} = -\frac{ie^2}{\hbar L^2} \sum_{\mathbf{k}} \sum_{n} f(E_{n\mathbf{k}}) \sum_{m(\neq n)} \left[ \frac{\langle u_{n\mathbf{k}} | \frac{\partial H}{\partial k_x} | u_{m\mathbf{k}} \rangle \langle u_{m\mathbf{k}} | \frac{\partial H}{\partial k_y} | u_{n\mathbf{k}} \rangle - \text{c.c.}}{(E_{n\mathbf{k}} - E_{m\mathbf{k}})^2} \right], \tag{2.116}$$

where $f(E_{n\mathbf{k}})$ is the Fermi distribution function, $L^2$ is the size of system, $|u_{n\mathbf{k}}\rangle$ is the Bloch wavefunction of the $n$th band with Bloch wavevector $\mathbf{k}$, $H$ is the Hamiltonian, and $E_{n\mathbf{k}}$ is the corresponding energy eigenvalue. Since the rightmost terms are precisely equal to the Berry curvature in Eq. (2.39), at zero temperature, it is hence rewritten in terms of the Chern number, also known as the TKNN (Thouless, Kohmoto, Nightingale, and den Nijs) number [14], associated with the $n$th band $C_n$:

$$\sigma_{xy} = -\frac{e^2}{h} \sum_{n \in \text{occ}} C_n, \quad C_n = \frac{1}{2\pi} \int_{\text{BZ}} d^2k \mathcal{F}_{n,xy}(\mathbf{k}), \tag{2.117}$$

where $\mathcal{F}_{n,xy}(\mathbf{k})$ is the Berry curvature related to the $n$th band given by Eq. (2.39). Obviously, the $n$th Chern number is always an integer, $C_n \in \mathbb{Z}$, and the total Chern number is equivalent to $C$ in Eq. (2.45).

The semial model with a nonzero Chern number realizing a Chern insulator was the honeycomb-lattice model in a periodic magnetic field proposed by Haldane [11]. Instead of the Haldane model, here we analyze the Wilson-Dirac model with a square lattice [15], as an example of the Chern insulator, described by

$$H = \sum_i \left[ \psi_{i+\hat{x}}^\dagger \frac{\sigma_z + i\sigma_x}{2} \psi_i + \psi_{i+\hat{y}}^\dagger \frac{\sigma_z + i\sigma_y}{2} \psi_i + \text{h.c.} \right] + m \sum_i \psi_i^\dagger \sigma_z \psi_i, \tag{2.118}$$

and the Bloch Hamiltonian is

$$\mathcal{H}(k_x, k_y) = \sin k_x \sigma_x + \sin k_y \sigma_y + \left(m + \cos k_x + \cos k_y\right) \sigma_z, \tag{2.119}$$

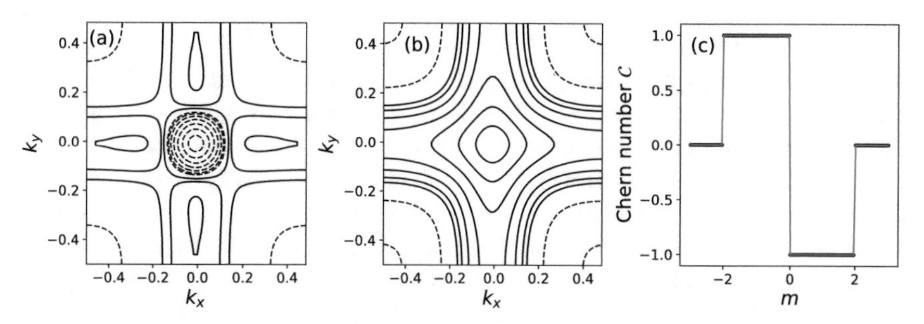


**Fig. 2.3** Contour plots of Berry curvature tr$\mathcal{F}(\mathbf{k})$ for the tight-binding model for the Chern insulator of Eq. (2.118) with **a** $m = -2.5$ and **b** $m = -1.5$. **a**, **b** Full and dashed lines denote positive and negative contour levels, respectively. Changing a total Berry flux of $2\pi$ by the band inversion gives a change of the Chern number by one between $m = -2.5$ and $m = -1.5$. **c** The numerical calculation (black dots) and the analytical solution from Eq. (2.122) (blue lines) for the Chern number $C$ with respect to $m$. The result that the Chern number is changed at $m = -2$, $m = 0$, and $m = 2$ totally agrees with the analytic result of Eq. (2.122)

where $\boldsymbol{\sigma}$ are Pauli matrices. The energy dispersion is easily obtained: $E_\pm(\mathbf{k}) = \epsilon(\mathbf{k}) \pm |\mathbf{d}(\mathbf{k})|$. The vector $\mathbf{d}(\mathbf{k})$ behaves as an effective Zeeman field applied to a pseudospin $\boldsymbol{\sigma}$ of a two-level system. In the present model, the vector $\mathbf{d}(\mathbf{k})$ is given by

$$\mathbf{d}(k_x, k_y) = \begin{pmatrix} d_x(\mathbf{k}) \\ d_y(\mathbf{k}) \\ d_z(\mathbf{k}) \end{pmatrix} = \begin{pmatrix} \sin k_x \\ \sin k_y \\ m + \cos k_x + \cos k_y \end{pmatrix}. \tag{2.120}$$

For simplicity, we take $\epsilon(\mathbf{k}) = 0$ and set the Fermi energy $E_F$ to be zero. From the energy dispersion $E_\pm = \pm\sqrt{(m + \cos k_x + \cos k_y)^2 + \sin^2 k_x + \sin^2 k_y}$, the band gap closes at $m = \pm 2, 0$. The gap closes at $(k_x, k_y) = (\pm\pi, \pm\pi)$ for $m = 2$, $(0, 0)$ for $m = -2$, and $(\pm\pi, 0)$, $(0, \pm\pi)$ for $m = 0$, respectively. According to Eq. (2.51), the Chern number is equal to the winding number of $\mathbf{d}(\mathbf{k})$ around the origin:

$$C = -\int_{\mathrm{BZ}} \frac{d^2k}{4\pi} \hat{\mathbf{d}} \cdot \left( \frac{\partial \hat{\mathbf{d}}}{\partial k_x} \times \frac{\partial \hat{\mathbf{d}}}{\partial k_y} \right). \tag{2.121}$$

Thus, the corresponding Chern number of this model is analytically calculated as

$$C = \begin{cases} 1 & \text{for } -2 < m < 0 \\ -1 & \text{for } 0 < m < 2 \\ 0 & \text{otherwise} \end{cases}. \tag{2.122}$$

This simple result gives us an insight into topological aspects and an intuitive understanding of quantum Hall states in band insulators.

A fundamental consequence of the topological classification of band insulators is the existence of gapless conducting states at interfaces between two systems with different values of the topological invariant. Such states appear as edge states at the interface between the Chern insulator and vacuum: this is well known as bulk-boundary correspondence. The existence of such edge states is deeply related to the topology of the corresponding bulk system. Suppose an interface where a crystal slowly interpolates between a Chern insulator ($C = 1$) and a trivial insulator ($C = 0$). Somewhere between the two insulators a gapless state has to appear, since otherwise it is impossible for the topological invariant to change. In the current model, one can compute the edge states using a simple theory of the Dirac model with a continuum limit. Consider an interface where the mass $m$ at one of the Dirac points changes sign, where the Hamiltonian near the Dirac point is described by $\mathcal{H} = k_x\sigma_x + k_y\sigma_y + m\sigma_z$. Let $y$ be a coordinate perpendicular to interface. The two regions $y > 0$ and $y < 0$ represent the trivial insulator and the Chern insulator, respectively. We then have $m \to m(y)$, where $m(y) < 0$ gives the Chern insulator for $y < 0$ and $m(y) > 0$ gives the trivial insulator for $y > 0$. Because $k_y$ is not a good quantum number now, by replacing $k_y \to -i\frac{\partial}{\partial y}$, our current model reduces to the (2+1)D massive Dirac Hamiltonian,

$$\mathcal{H} = k_x\sigma_x - i\frac{\partial}{\partial y}\sigma_y + m(y)\sigma_z. \tag{2.123}$$

Since a state $\exp\left[-\int_0^y dy' m(y')\right]\ {}^t(1, -1)$ is an eigenstate with zero energy of the Hamiltonian Eq. (2.123), a state

$$\psi(x, y) = \frac{N}{\sqrt{2}} e^{ik_x x} e^{-\int_0^y dy' m(y')} \begin{pmatrix} 1 \\ -1 \end{pmatrix}, \tag{2.124}$$

corresponds to an eigenstate with $E = k_x$. Here $N$ is a normalization constant. Physically, this eigenstate intersects the Fermi energy $E_F$ with a positive group velocity $dE/dk_x > 0$ and describes a right-going chiral edge state. This simple and elegant exact solution implies that the topology appears as low-energy edge states living near the system edges as well as localized states near impurities. This interplay between topology and gapless modes is ubiquitous in physics and appears in many contexts. The bulk-boundary correspondence for the quantum Hall systems is explained by means of the Laughlin's gedanken experiment [16]. In electronic systems, this correspondence can be described in terms of polarization in the solid state physics associated with the flow of hybrid Wannier centers [17–19]. Namely, as we stated in Sect. 2.1.4, the Chern number corresponds to the flow of hybrid Wannier centers via modern theory of polarization [17–19].

We demonstrate the above argument by performing numerical calculations of Eq. (2.118). We have used the open-source PythTB [2]. Figure 2.2a–c show the bulk band structure of Eq. (2.118) at (a) $m = -2.5$, (b) $m = -2$, and (c) $m = -1.5$. If we set the Fermi energy to be zero, one can find the conduction band and the valence band at $\Gamma$ are inverted between $m < -2$ (Fig. 2.2a) and $m > -2$ (Fig. 2.2c). We can also check the Berry curvature and the Chern number. Figure 2.3a, b show a

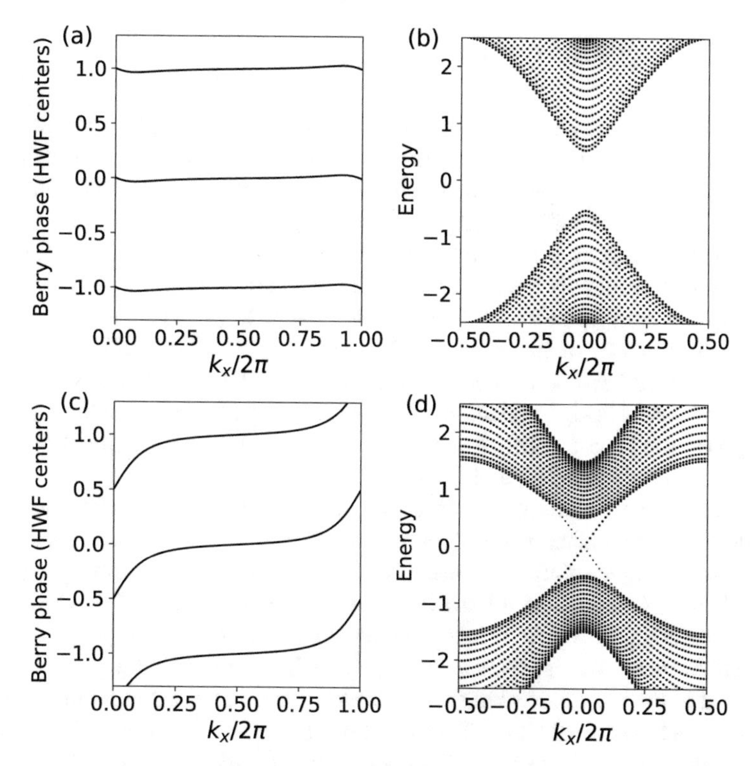


**Fig. 2.4** Hybrid Wannier centers (Berry phase) indicating bulk topologies and edge band structures on a ribbon cut from the tight-binding model for the Chern insulator of Eq. (2.118) with **a**, **b** $m = -2.5$ and **c**, **d** $m = -1.5$, respectively. **a** Gapped hybrid Wannier centers correspond to **b** the absence of edge states in the trivial phase, whereas **c** gapless hybrid Wannier centers correspond to **d** the nontrivial edge states in the Chern insulator. Edge states on the top and bottom edges of the ribbon are indicated by full and reduced intensity, respectively

contour plot of the Berry curvature for the occupied lower band. The total Berry flux near $\Gamma$ is $-\pi$ at $m = -2.5$ while that is replaced by $+\pi$ at $m = -1.5$. Namely, between $m = -2.5$ and $m = -1.5$, the Chern number jumps from $C = 0$ to $C = 1$ revealed by the change of $2\pi$ for the total Berry flux. This can be also demonstrated by a direct calculation for the Chern number (Fig. 2.3) which is totally consistent with our analytic calculation in Eq. (2.122). Furthermore, the Berry phases (from the flow of hybrid Wannier centers) in bulk and edge band structures for a ribbon geometry along $x$ direction are shown in Fig. 2.4. In the trivial case of $m = -2.5$, the hybrid Wannier centers are gapped in Fig. 2.4a and there are no gapless edge states in Fig. 2.4b. For the Chern insulator case of $m = -1.5$, the hybrid Wannier centers shift up by one unit as $k_x$ cycles in Fig. 2.4c, corresponding to the Chern number $C$ to be one. Therefore, we can confirm the gapless chiral surface states in Fig. 2.4d.

### 2.3.2 Topological Insulators

Since the Hall conductivity is odd under time-reversal operation, the Chern insulator requires breaking of time-reversal symmetry. However, effects of electron spins have been ignored here. The spin-orbit coupling allows a different topological insulating system in the presence of time-reversal symmetry. This novel topological phase of insulating band structures is called $Z_2$ topological insulators (TIs) [20, 21]. Historically, spin Hall insulators have been proposed in an effort to understand an intrinsic mechanism of spin Hall effect for spinful particles [22, 23]. This proposal has triggered realizations of quantum spin Hall insulators (2D TIs) and 3D TIs. So far, various TIs have been confirmed by experimentally. The 2D $Z_2$ TIs were experimentally observed in HgTe quantum wells [24] based on a theoretical model [25]. In 3D, $Z_2$ TIs have been confirmed in various materials such as $Bi_{1-x}Sb_x$ [26], $Bi_2Se_3$ family of materials [27–29], and so on.

First of all, we define the $Z_2$ topological invariant which determines whether a 2D bulk insulating system is the TI or not. The key concept to grasp TIs is the time-reversal symmetry in spinful systems. It is well-known that the time-reversal operator $\Theta$ is written as $\Theta = K$ in spinless systems, and $\Theta = i\sigma_y K$ in spinful systems from quantum mechanics, where $K$ is a complex conjugate and $\sigma_y$ is a $y$-component of the Pauli matrices. We consider spinful systems in the following argument. As we mentioned in Sect. 2.2.1, if a bulk insulating system is invariant under time-reversal symmetry, the energies of the Bloch functions at $\mathbf{k}$ and $-\mathbf{k}$ are equal. It is called Kramers' degeneracy. The $\mathbf{k}$ vector satisfying $\mathbf{k} \equiv -\mathbf{k} \pmod{\mathbf{G}}$ in momentum space where $\mathbf{G}$ is a reciprocal vector is called time-reversal invariant momentum (TRIM). In 2D systems, there exist four TRIMs (Fig. 2.5a)

$$\mathbf{\Gamma}_{i=n_1,n_2} = \frac{1}{2}(n_1\mathbf{G}_1 + n_2\mathbf{G}_2) \quad (n_1, n_2 = 0, 1), \tag{2.125}$$

where $\mathbf{G}_{1,2}$ are the primitive reciprocal lattice vectors. Moreover, as in the similar manner in Sect. 2.2.1, we define a $\mathrm{U}(2N)$ matrix

$$w_{mn}(\mathbf{k}) = \langle u_{m,-\mathbf{k}}|\Theta|u_{n\mathbf{k}}\rangle, \tag{2.126}$$

where $|u_{n\mathbf{k}}\rangle$ $(n = 1, \dots, 2N)$ represent occupied states below the Fermi energy. For simplicity, we assume that there are no degeneracies other than those required by time-reversal symmetry. Then, the $2N$ eigenstates may be divided into $N$ pairs of the form $\{|u_{n,-\mathbf{k}}\rangle, \Theta|u_{n\mathbf{k}}\rangle\}$. With this choice of eigenstates, $w_{mn}$ is a block-diagonal matrix with $2 \times 2$ matrices on the diagonal blocks, and in addition, $w_{mn}$ becomes *antisymmetric* at TRIMs. An antisymmetric matrix is characterized by its Pfaffian, whose square is equal to the determinant. Hence at each TRIM we can find a quantity only taking $\pm 1$ as

$$\delta_i \equiv \frac{\mathrm{Pf}[w(\Gamma_i)]}{\sqrt{\det[w(\Gamma_i)]}}, \tag{2.127}$$

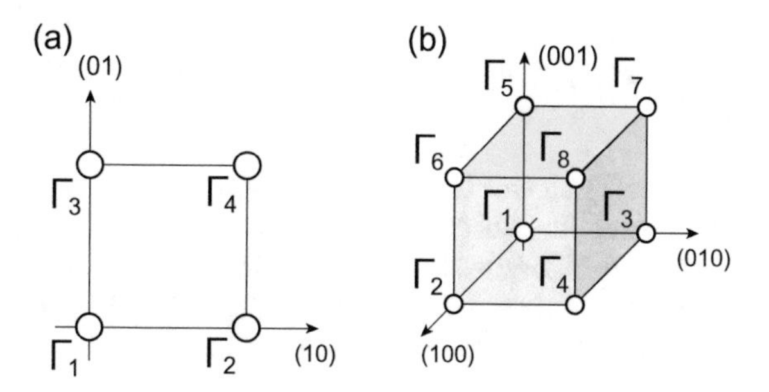


**Fig. 2.5** Time-reversal-invariant momenta (TRIMs). **a** There are four TRIMs in a two-dimensional Brillouin zone and **b** there are eight TRIMs for a three-dimensional Brillouin zone

because $(\mathrm{Pf}[w])^2 = \det[w]$. Therefore, the $Z_2$ topological invariant $\nu$ (mod 2) for 2D TIs can be defined using $\delta_i$ as [30]

$$(-1)^\nu = \prod_{i \in \mathrm{TRIM}} \delta_i. \tag{2.128}$$

One can easily see that the $Z_2$ topological invariant $\nu$ should be either 0 or 1, defined modulo 2. From Eq. (2.128), it seems that $\nu$ can be calculated from the wavefunctions *only* at the TRIM. However, the ambiguity in choosing the branches of $\sqrt{\det[w(\Gamma_i)]}$ at $k = \mathbf{\Gamma}_i$ affects the values of $\delta_i (= \pm 1)$. Here, in fact one should evaluate Eq. (2.128) to eliminate the ambiguity by taking the branch chosen continuously over a half of the BZ, and it is equally given by

$$\nu = \frac{1}{2\pi} \left[ \oint_{\partial \mathrm{BZ}'} dl \ \mathrm{tr}\mathcal{A} - \int_{\mathrm{BZ}'} d^2k \ \mathrm{tr}\mathcal{F} \right], \tag{2.129}$$

where BZ$'$ denotes the half BZ ($k_x \in [0, \pi]$, $k_y \in [0, 2\pi]$) and $\partial$BZ$'$ is its boundary [30]. This is known as the obstruction formulation [30].

In Sect. 2.1.4, we have shown that the flow of hybrid Wannier centers gives the Chern number and it is explained by polarization theory [17–19]. Similarly, we can find the expression of the $Z_2$ topological invariant in terms of the flow of hybrid Wannier centers [31–33] by introducing time-reversal polarization [30]. Here we briefly explain how to connect between the $Z_2$ topological invariant and the Wilson loop which corresponds to hybrid Wannier centers (Eq. (2.61)). Let us consider the U($2N$) Wilson loop along the same constant $k_y$ loops:

$$W(k_y) = \mathcal{P} \left[ i \oint dk_x \mathcal{A}_x(k_x, k_y) \right] \in \mathrm{U}(2N). \tag{2.130}$$

As we stated in Sect. 2.1.4, since the U(1) eigenvalue of the Wilson loop corresponds to the U(1) gauge part of the Berry connection, we recast $\nu$ in Eq. (2.129) into [33]

$$\nu = \frac{1}{2\pi}\left(\sum_n \int_0^{\pi} dk_y \partial_{k_y} \phi_n(k_y) - \sum_n [\phi_n(\pi) - \phi_n(0)]\right) \pmod 2, \quad (2.131)$$

where $\phi_n(k_y)$ is the U(1) eigenvalues of the Wilson loop $\det W(k_y) = \exp(i \sum_n \phi_n(k_y))$ and we have used the Stokes' theorem. Moreover, $e^{i\phi_n(k_y)}$ is doubly degenerate at $k_y = 0$ and $k_y = \pi$ because Eq. (2.83) indicates $W(-k_y) = UW^T(k_y)U^{-1}$ with $U$ an antisymmetric matrix which does not depend on $\mathbf{k}$. Thus, by choosing $\phi_n(0)$ and $\phi_n(\pi)$ to be in $[0, 2\pi)$, we rewritten Eq. (2.131) as

$$\nu = \sum_n N_n \pmod 2, \quad (2.132)$$

where $N_n$ is the winding number of the phase $\phi_n$ [33]. The number $\sum_n N_n$ is simply counted by tracking the hybrid Wannier functions for the eigenvalues $\phi_n$ [3].

We now proceed to define $Z_2$ topological invariants for 3D TIs in the same manner. There are eight TRIMs in 3D BZ (Fig. 2.5b) written as

$$\Gamma_{i=n_1,n_2,n_3} = \frac{1}{2}(n_1\mathbf{G}_1 + n_2\mathbf{G}_2 + n_3\mathbf{G}_3) \quad (n_1, n_2, n_3 = 0, 1). \quad (2.133)$$

The 3D TIs are characterized by four $Z_2$ topological invariants $\nu_0$; $(\nu_1\nu_2\nu_3)$ defined as

$$(-1)^{\nu_0} = \prod_{i=1}^{8} \delta_i, \quad (-1)^{\nu_{k=1,2,3}} = \prod_{n_k=1;n_{j\neq k}=0,1} \delta_{i=(n_1n_2n_3)}. \quad (2.134)$$

They are defined as products of the indices at some of TRIM. Because of the gauge degree of freedom, these four products are the only combinations which are independent and gauge invariant. The phases with $\nu_0 = 1$ are called strong TIs (STIs), those with $\nu_0 = \nu_1 = \nu_2 = \nu_3 = 0$ are trivial insulators, and the others are weak TIs (WTIs). Difference between the STI and the WTI lies in robustness of their topological states. The indices $(\nu_1\nu_2\nu_3)$ for the WTI indicate Miller indices, along which the system is regarded as layered 2D TIs. On the other hand, $\nu_0$ implies the number of band inversions in the whole BZ [30]. If band inversions occur odd/even times, the system is topologically nontrivial/trivial. In this case, surface Fermi surfaces in the STI/WTI enclose the odd/even number of surface TRIM. Therefore, the STI phase is robust against disorder and perturbations preserving TRS, because the system is topologically nontrivial. In contrast the WTI phase is not robust.

In the presence of additional inversion symmetry, Kramers' degeneracy appears at every $\mathbf{k}$ point. Therefore, we obtain

$$\delta_i = \prod_{n\in\mathrm{occ}} \xi_{2n}(\mathbf{\Gamma}_i), \quad (2.135)$$

where $\xi_n(\mathbf{\Gamma}_i)$ is an inversion parity for the $n$th wavefunction at the TRIM $\mathbf{\Gamma}_i$ and the product is taken over the occupied states. Note that $\xi_{2n-1} = \xi_{2n}$ due to Kramers' degeneracy. Hence, the change of the $Z_2$ topological invariant implies that once the band gap closes and reopens at the TRIM $\mathbf{\Gamma}_i$.

Consequently, bulk-boundary correspondence allows Dirac cones, which are topologically nontrivial surface states for the TI. The surface states of the TI have unique and exotic properties which are absent in those of quantum Hall insulators or in Dirac cones of graphene. Firstly, the surface states for the TI are robust against impurities or disorders unless TRS is broken. It is exactly a statement of bulk-boundary correspondence that topological properties of surface states are determined by the bulk physics. These surface states have fascinating features that they have spin-polarized Fermi surfaces and they carry spin polarized current. It can be understood intuitively since the TI (a quantum spin Hall system) can be regarded as a superposition of quantum Hall systems with spin-up ($s_z = +1/2$) electrons and spin-down ($s_z = -1/2$) electrons with opposite Chern numbers. Therefore, back scattering from $\mathbf{k}$ to $-\mathbf{k}$ is prohibited because of spin-polarized Fermi surface.

### *2.3.3 Topological Crystalline Insulators*

Beyond internal symmetries, crystal symmetries of crystalline materials, such as spatial inversion, mirror, and rotational symmetries, play a crucial role to understand such topological phases. Topological crystalline insulators (TCIs) are topological phases of matter that are ensured by those crystal symmetries. TCIs govern topological properties such as robust conducting surface states as long as the focused crystal symmetry is preserved. Given the richness and complexity of crystallography, since the seminal proposal of a TCI protected by $C_4$ rotational symmetry [34], several theoretical works proposed TCI phases ensured by inversion symmetry [35], rotational symmetry [6, 36, 37], mirror symmetry [38], and candidate materials include rocksalt semiconductors [39], pyrochlores [40], graphene multilayers [41], heavy fermion systems [42], and antiperovskites [43]. Thus, various combinations of internal symmetry and crystal symmetries have turned out to give an immense list of new topological phases [44–49]. In particular, mirror-symmetric TCI phases have attracted interest because they present the first material realization in IV-VI semiconductors SnTe, $Pb_{1-x}Sn_xTe$, and $Pb_{1-x}Sn_xSe$. There TCI phases are supported by the ARPES (angle-resolved photoemission spectroscopy) experiments in the same materials [38, 50–55], and analyzed in numerous works [56–61].

We introduce, as an example, basic properties of TCIs of the SnTe class protected by mirror symmetry. Assume that the system preserves mirror symmetry with respect to the $xy$ plane. Then, the Hamiltonian satisfies

$$M\mathcal{H}(k_x, k_y, k_z)M^{-1} = \mathcal{H}(k_x, k_y, -k_z), \tag{2.136}$$

where $M$ is the mirror operator. In spin-orbit-coupled systems, the mirror operator $M$ satisfies $M^2 = -1$, and therefore, eigenvalues of $M$ are either $+i$ or $-i$. This implies that on the mirror plane (hereafter denoted by $k_z = 0$) energy eigenstates can be chosen to be eigenstates of $M$ because the Hamiltonian commutes with $M$. Thus the eigenstates are divided into two subspaces by the mirror eigenvalues:

$$\mathcal{H}(k_x, k_y, k_z = 0) = \mathcal{H}_{+i}(k_x, k_y, k_z = 0) \oplus \mathcal{H}_{-i}(k_x, k_y, k_z = 0). \tag{2.137}$$

A similar formula holds for the $k_z = \pi$ plane. For each subspace of eigenstates with the mirror eigenvalue $\eta$, a corresponding Chern number $C_\eta$ $(\eta = \pm i)$ can be defined. TRS requires that the total Chern number $C = C_{+i} + C_{-i}$ related to the quantized Hall conductivity is zero. On the other hand, a new topological invariant called *mirror Chern number* [62],

$$C_M \equiv \frac{C_{+i} - C_{-i}}{2}, \tag{2.138}$$

can be a nonzero integer. Hence, 3D TCI phases with mirror symmetry are classified by mirror Chern numbers associated with mirror-invariant planes in momentum space.

Besides symmorphic symmetries, topological phases with nonsymmorphic symmetries have been vigorously discussed [63–68]. From the proposal of a nonsymmorphic TCI [69], glide symmetry, which is a combined symmetry between a reflection operation and a half translation operation, is particularly recognized to ensure a rich variety of TCI phases, such as a $Z_2$-characterized TCI with the Möbius twist single Dirac cone in surface states in the absence of time-reversal symmetry [70–72], and that with hourglass-like surface states which are confirmed in KHg$X$ ($X$=As, Sb, Bi) [73–77]. A key for realization of a TCI protected by nonsymmorphic symmetries is additional band degeneracy in nonsymmorphic systems. Since nonsymmorphic symmetry contains products of point-group operation and a nonprimitive translation, the periodicity of wavefunctions is different from that under symmorphic symmetries.

So far, many previous works for searching topological materials focused on time-reversal invariant systems with significant spin-orbit coupling in the context of the $Z_2$ TI. A recent breakthrough of the realization for TCIs however occurs in the $MnBi_2Te_{3n+1}$ family of materials exhibiting an antiferromagnetic TI [78–84]. These suggestions have broadened a variety of topological materials including the glide-$Z_2$ TCI phase.

Very recently, a new class of TCIs hosting gapless boundary states whose dimension is less than $d - 1$ for a $d$D insulating bulk has also been established, and they are dubbed *higher-order TIs* [85–94]. Theoretical works support existence of the higher-order TI by topological classifications and model constructions, and they have also proposed a few candidates materials [95–99]. The only evidence for electronic higher-order TIs was experimentally investigated in bismuth [95]. Such a higher-order TI can be characterized by symmetry-based indicators. Nonetheless, the explanation of bulk-hinge correspondence and the interplay of symmetry still remain to be explored.

For comprehensive classifications of topological phases protected by a space group, an approach based on the $K$-theory, which is a mathematical approach for gapped systems, has been adopted [10, 100–103]. Other approaches, such as topological quantum chemistry and symmetry-based indicators, have been attracting research interest [7, 8, 104, 105]; in these approaches, topology of the bands and their compatibility relations are studied in the context of topological phases protected by space-group symmetries. Particularly, symmetry-based indicators are useful in diagnosing topologically distinct band structures from the combinations of irreducible representations at high-symmetry momenta [7, 8]. These approaches focus on different aspects of topological phases. In the $K$-theory approach, one can comprehensively classify nontrivial phases, whereas an explicit formula for the topological invariant does not follow immediately from the theory. On the other hand, the symmetry-based indicator can reveal only the topological phases characterized by combinations of irreps. Thus, this theory cannot capture topological phases which cannot be known only from the irreps. Thus even with these powerful tools, one cannot reach a full understanding of nontrivial topological phases protected by space-group symmetries, and there is much room for further investigation.

### *2.3.4 Topological Semimetals; Weyl, Dirac, and Nodal-Line Semimetals*

In general, a band gap in crystals closes usually at high-symmetry points, originating from higher-dimensional irreducible representation of the $k$-group, and at generic $\mathbf{k}$ points, the bands anticross by level repulsion. Nevertheless, the band gap may close at generic points by topological reasons; we call this topological semimetals. The topological semimetals are generally classified into topological nodal-point semimetals with point nodes and topological nodal-line semimetals with zero-band-gap line nodes. At these point nodes and line nodes, the band gap closes [106].

#### 2.3.4.1 Weyl Semimetal

In *Weyl semimetals* (WSMs) [107–110], conduction and valence bands touch only at discrete points at or around the Fermi level in bulk, and they disperse linearly in all directions around these critical points in momentum space where all the bands are non-degenerate. The vertex of the gapless points is called a Weyl node [107, 111], generally described by

$$\mathcal{H}_{\mathrm{Weyl}}(\mathbf{k}) = v\mathbf{k} \cdot \boldsymbol{\sigma}, \tag{2.139}$$

where $v$ is a constant and $\boldsymbol{\sigma} = (\sigma_x, \sigma_y, \sigma_z)$. Each Weyl node is characterized by the monopole density in Eq. (2.40). Namely, the 3D Weyl nodes behave either as a source (a monopole) or a drain (an antimonopole) for the Berry curvature $\mathcal{F}(\mathbf{k})$. One can easily see that $\rho_n(\mathbf{k})$ vanishes identically unless the $n$th band touches another band, where the monopole density has a $\delta$-function singularity. Furthermore the coefficient of the $\delta$-function in the monopole density should be an integer in the unit of $2\pi$, which is called a monopole charge. As we mentioned above, the 3D Weyl nodes are either a source (charge $+1$) or a drain (charge $-1$). Since the monopole charge is quantized, the only chances to change the monopole charge are pair creations or pair annihilations, under some change of a parameter contained in the Hamiltonian. The 3D Weyl nodes hence are stable unless interactions or perturbations breaking translation symmetry are applied. The robustness of 3D Weyl nodes can be quantified by the Chern number of the valence band on a sphere surrounding the point, since this Chern number (2.51) is equal to the monopole charge (2.40).

An intriguing feature of Weyl nodes entails the appearance of a surface Fermi surface forming an open arc, so-called *Fermi arc* [110, 112–114], even though usual Fermi surfaces form closed loops in solid state physics. The reason of the existence of a Fermi arc is the topological property of Weyl nodes, which are projected into end points of the Fermi arc. Suppose a Weyl node is located at $\mathbf{k}^{\mathrm{W}} = (k_x^{\mathrm{W}}, k_y^{\mathrm{W}}, k_z^{\mathrm{W}})$. When we slice a 2D plane $k_z = k_z^0$ in the 3D BZ, we can calculated the Chern number on the sliced plane $C(k_z^0)$. Then, the difference between the Chern number at $k_z^{\mathrm{W}} + \delta$ and that at $k_z^{\mathrm{W}} - \delta$ is

$$C(k_z^{\mathrm{W}} + \delta) - C(k_z^{\mathrm{W}} - \delta) = \frac{1}{2\pi} \sum_{n\in\mathrm{occ}} \int_S dS \mathcal{F}_n(\mathbf{k}) \cdot \mathbf{n}, \tag{2.140}$$

where $\delta$ is a small positive constant, $S$ represents two planes $k_z = k_z^{\mathrm{W}} \pm \delta$, and $\mathbf{n}$ is a normal vector of the surface $S$ (Fig. 2.6). Recall that on a closed surface the integral of the Berry curvature is always an integer (Sect. 2.1.3). By reducing the radius of the sphere to zero, it reduces to the monopole charge of the Weyl node enclosed. Namely, when $k_z$ is changed across the monopole (i.e., Weyl nodes with $\rho = +1$), the Chern number changes by $+1$, and when $k_z$ is changed across an antimonopole (i.e., Weyl nodes with $\rho = -1$), it is changed by $-1$. Therefore, the Chern number in the region $k_z > k_z^{\mathrm{W}}$ is given by $+1$ and there exists a chiral edge state, whereas a chiral edge state is absent in $k_z < k_z^{\mathrm{W}}$ due to the Chern number being zero. This is the nature of the appearance of the Fermi arc (Fig. 2.6).

It has been known that the WSM phase appears when either inversion or time-reversal symmetry is broken, because the Berry curvature is identically zero under both of inversion and time-reversal symmetries in any momentum $\mathbf{k}$. As we have shown earlier, the Berry curvature is even under inversion symmetry and odd under time-reversal symmetry. Therefore, inversion and time-reversal symmetries imply $\rho_n(\mathbf{k}) = -\rho_{n_\mathcal{I}}(-\mathbf{k})$ and $\rho_n(\mathbf{k}) = \rho_{n_\mathcal{T}}(-\mathbf{k})$, where $n_\mathcal{I}$ and $n_\mathcal{T}$ are band indices which are obtained from the inversion and time-reversal operations on the $n$th band, respectively.

Several works have proposed noncentrosymmetric WSMs in $\mathrm{LaBi_{1-x}Sb_xTe_3}$, $\mathrm{LuBi_{1-x}Sb_xTe_3}$ [115], transition-metal dichalcogenides [116, 117], Te at high pressure [118], and $\mathrm{SrSi_2}$ [119], and those WSMs are established experimentally in TaAs

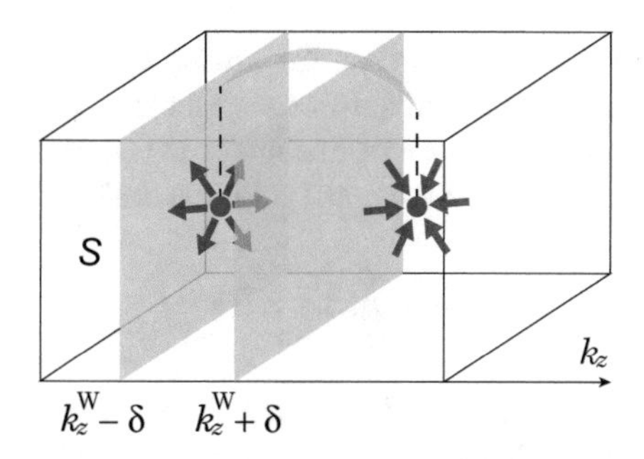


**Fig. 2.6** A schematic figure of the Fermi arc. From bulk-boundary correspondence, the Fermi arc (yellow line) which are connecting the projections of bulk 3D Weyl nodes appear in the interfaces between the WSM and vacuum. The red arrows denote the Berry curvature, indicating that the two Weyl nodes are a monopole (source) and an antimonopole (drain)

family materials [120–124] as the first realistic materials of WSMs in accordance with theoretical predictions [125, 126]. In contrast, there have been pyrochlore iridates $A_2Ir_2O_7$ (A =Y, Eu, Sm, Nd) [110, 112], $HgCr_2Se_4$ [127], and Co-based magnetic Heusler alloys $XCo_2Sn$ (X = Zr, Nb, V) [128], but they have not been experimentally verified. Very recently, $Co_3Sn_2S_2$ turned out to host a magnetic WSM phase [129], which opens up new possibilities for topological materials.

#### 2.3.4.2 Dirac Semimetal

Dirac semimetals are phases with 3D Dirac nodes which have Kramers' degeneracy in momentum space [130]. Thus, inversion and time-reversal symmetries are required to realize Dirac semimetals. Because the 3D Dirac node is a superposition of two Weyl nodes with a monopole and an antimonopole, it is precarious because it may allow off-diagonal mass terms which gaps at the Dirac node. For this reason, additional crystal symmetry is needed to protect stable Dirac nodes' degeneracy [131]. In particular, rotational symmetry plays a crucial role in $A_3Bi$ (A=Na, K, Rb) [132] and $Cd_3As_2$ [133], and among those materials, Dirac nodes in $Na_3Bi$ [134] and $Cd_3As_2$ [135, 136] are measured experimentally.

#### 2.3.4.3 Nodal-Line Semimetal

In nodal-line semimetals, the gap closes not at isolated points but along line nodes in bulk. There are two typical mechanisms for closing of a gap along a line node: nodal lines stemming from (A) mirror (glide) symmetry and from (B) $\pi$ Berry phase.

In the case (A), a nodal line emerges on the mirror (glide) plane if the valence and conduction bands have opposite mirror (glide) eigenvalues. Numerous materials are regarded as this class of nodal-line semimetals such as carbon allotropes [137, 138],

$Ca_3P_2$ [139, 140], $Cu_3$(Pd, Zn)N [141, 142], and so on. Experimentally, ZiSiS has been confirmed to be this class of nodal-line semimetals [143–145].

Another reason of the appearance of line nodes in (B) occurs in spinless systems with inversion and time-reversal symmetries. In this case, the Berry phase

$$\varphi = \oint_l d\mathbf{k} \cdot \mathcal{A}_n(\mathbf{k}), \tag{2.141}$$

have to be quantized 0 or $\pi$ for any loop $l$ in $k$-space. If the loop $l$ encircles a line node, $\varphi$ is equal to $\pi$, and otherwise $\varphi$ is to be zero. Alkaline-earth compounds $AX_2$ (A = Ca, Sr, Ba; X = Si, Ge, Sn) [146] and alkaline-earth metals [147, 148] belong this class of nodal-line semimetals.

### *2.3.5 Topological Photonics and Other Topological Phases*

We have introduced topological phases in the context of the electronic band theory for a single-particle. The concept of topological matters has been broadened to a single quasi-particle system, such as a photonic crystal system or a magnonic system, and they are one of the best candidate systems for realizing these physics.

Photonic crystals yield Bloch-like states (Sect. 2.1.5) and have been explored with the rapid growth of interest in topological phases from the similarity with the Bloch electrons. The seminal proposals for 2D quantum Hall states in photonic crystals pointed out that topological band structures are indeed a ubiquitous property of waves inside a periodic medium regardless of the classical or quantum nature of the waves [149, 150], and those are observed in a realistic model [151, 152]. After that, several works have proposed a direct analogue to a 2D $Z_2$ TI with opposite chiralities for edge states [153–158], a 3D $Z_2$ TI [159, 160], Weyl and line nodes [161], a glide-symmetric $Z_2$ TCI [72], a $C_4$-symmetric TCI with time-reversal symmetry [162], a 3D Chern insulator [163] in photonic crystals, and so on.

The systems for magnons [164–174], phonons [175–187] or plasmons [188–190] are also good candidates for applying topological properties and other quasi-particle systems are vigorously studied these days. These fertile proposals beyond electronic systems yield exotic properties of bosons and possibilities towards application of topological materials.

## References

1. Berry MV (1984) Proceedings of the royal society of London A. Math Phys Sci 392(1802):45. https://doi.org/10.1098/rspa.1984.0023
2. Vanderbilt D (2018) Berry phases in electronic structure theory: electric polarization, orbital magnetization and topological insulators. Cambridge University Press

3. Gresch D, Autès G, Yazyev OV, Troyer M, Vanderbilt D, Bernevig BA, Soluyanov AA (2017) Phys Rev B 95:075146. https://doi.org/10.1103/PhysRevB.95.075146
4. Joannopoulos JD, Johnson SG, Winn JN, Meade RD (2008) Molding the flow of light
5. Altland A, Zirnbauer MR (1997) Phys Rev B 55:1142. https://doi.org/10.1103/PhysRevB.55.1142
6. Fang C, Gilbert MJ, Bernevig BA (2012) Phys Rev B 86:115112. https://doi.org/10.1103/PhysRevB.86.115112
7. Watanabe H, Po HC, Vishwanath A (2018) Sci Adv 4(8):eaat8685
8. Po HC, Vishwanath A, Watanabe H (2017) Nat Commun 8(1):50
9. Inui T, Tanabe Y, Onodera Y (2012) Group theory and its applications in physics, vol 78. Springer Science & Business Media
10. Shiozaki K, Sato M, Gomi K (2018) arXiv:1802.06694
11. Haldane FDM (1988) Phys Rev Lett 61:2015. https://doi.org/10.1103/PhysRevLett.61.2015
12. Chang CZ, Zhang J, Feng X, Shen J, Zhang Z, Guo M, Li K, Ou Y, Wei P, Wang LL, Ji ZQ, Feng Y, Ji S, Chen X, Jia J, Dai X, Fang Z, Zhang SC, He K, Wang Y, Lu L, Ma XC, Xue QK (2013) Science 340(6129):167. https://doi.org/10.1126/science.1234414
13. Jotzu G, Messer M, Desbuquois R, Lebrat M, Uehlinger T, Greif D, Esslinger T (2014) Nature 515(7526):237
14. Thouless DJ, Kohmoto M, Nightingale MP, den Nijs M (1982) Phys Rev Lett 49:405. https://doi.org/10.1103/PhysRevLett.49.405
15. Qi XL, Hughes TL, Zhang SC (2008) Phys Rev B 78:195424. https://doi.org/10.1103/PhysRevB.78.195424
16. Laughlin RB (1981) Phys Rev B 23:5632. https://doi.org/10.1103/PhysRevB.23.5632
17. Resta R (1992) Ferroelectrics 136(1):51
18. King-Smith RD, Vanderbilt D (1993) Phys Rev B 47:1651. https://doi.org/10.1103/PhysRevB.47.1651
19. Vanderbilt D, King-Smith RD (1993) Phys Rev B 48:4442. https://doi.org/10.1103/PhysRevB.48.4442
20. Hasan MZ, Kane CL (2010) Rev Mod Phys 82:3045. https://doi.org/10.1103/RevModPhys.82.3045
21. Qi XL, Zhang SC (2011) Rev Mod Phys 83:1057. https://doi.org/10.1103/RevModPhys.83.1057
22. Murakami S, Nagaosa N, Zhang SC (2003) Science 301(5638):1348. https://doi.org/10.1126/science.1087128
23. Sinova J, Culcer D, Niu Q, Sinitsyn NA, Jungwirth T, MacDonald AH (2004) Phys Rev Lett 92:126603. https://doi.org/10.1103/PhysRevLett.92.126603
24. König M, Wiedmann S, Brüne C, Roth A, Buhmann H, Molenkamp LW, Qi XL, Zhang SC (2007) Science 318(5851):766. https://doi.org/10.1126/science.1148047
25. Bernevig BA, Hughes TL, Zhang SC (2006) Science 314(5806):1757. https://doi.org/10.1126/science.1133734
26. Hsieh D, Qian D, Wray L, Xia Y, Hor YS, Cava RJ, Hasan MZ (2008) Nature 452(7190):970
27. Zhang H, Liu CX, Qi XL, Dai X, Fang Z, Zhang SC (2009) Nat Phys 5(6):438
28. Xia Y, Qian D, Hsieh D, Wray L, Pal A, Lin H, Bansil A, Grauer D, Hor YS, Cava RJ et al (2009) Nat Phys 5(6):398
29. Chen YL, Analytis JG, Chu JH, Liu ZK, Mo SK, Qi XL, Zhang HJ, Lu DH, Dai X, Fang Z, Zhang SC, Fisher IR, Hussain Z, Shen ZX (2009) Science 325(5937):178. https://doi.org/10.1126/science.1173034
30. Fu L, Kane CL (2006) Phys Rev B 74:195312. https://doi.org/10.1103/PhysRevB.74.195312
31. Soluyanov AA, Vanderbilt D (2011) Phys Rev B 83:035108. https://doi.org/10.1103/PhysRevB.83.035108
32. Soluyanov AA, Vanderbilt D (2011) Phys Rev B 83:235401. https://doi.org/10.1103/PhysRevB.83.235401
33. Yu R, Qi XL, Bernevig A, Fang Z, Dai X (2011) Phys Rev B 84:075119. https://doi.org/10.1103/PhysRevB.84.075119

34. Fu L (2011) Phys Rev Lett 106:106802. https://doi.org/10.1103/PhysRevLett.106.106802
35. Hughes TL, Prodan E, Bernevig BA (2011) Phys Rev B 83:245132. https://doi.org/10.1103/PhysRevB.83.245132
36. Fang C, Gilbert MJ, Bernevig BA (2013) Phys Rev B 87:035119. https://doi.org/10.1103/PhysRevB.87.035119
37. Alexandradinata A, Fang C, Gilbert MJ, Bernevig BA (2014) Phys Rev Lett 113:116403. https://doi.org/10.1103/PhysRevLett.113.116403
38. Hsieh TH, Lin H, Liu J, Duan W, Bansil A, Fu L (2012) Nat Commun 3:982. https://doi.org/10.1038/ncomms1969
39. Sun Y, Zhong Z, Shirakawa T, Franchini C, Li D, Li Y, Yunoki S, Chen XQ (2013) Phys Rev B 88:235122. https://doi.org/10.1103/PhysRevB.88.235122
40. Kargarian M, Fiete GA (2013) Phys Rev Lett 110:156403. https://doi.org/10.1103/PhysRevLett.110.156403
41. Kindermann M (2015) Phys Rev Lett 114:226802. https://doi.org/10.1103/PhysRevLett.114.226802
42. Weng H, Zhao J, Wang Z, Fang Z, Dai X (2014) Phys Rev Lett 112:016403. https://doi.org/10.1103/PhysRevLett.112.016403
43. Hsieh TH, Liu J, Fu L (2014) Phys Rev B 90:081112. https://doi.org/10.1103/PhysRevB.90.081112
44. Slager RJ, Mesaros A, Juricic V, Zaanen J (2013) Nat Phys 9:98. https://doi.org/10.1038/nphys2513
45. Chiu CK, Yao H, Ryu S (2013) Phys Rev B 88:075142. https://doi.org/10.1103/PhysRevB.88.075142
46. Morimoto T, Furusaki A (2013) Phys Rev B 88:125129. https://doi.org/10.1103/PhysRevB.88.125129
47. Jadaun P, Xiao D, Niu Q, Banerjee SK (2013) Phys Rev B 88:085110. https://doi.org/10.1103/PhysRevB.88.085110
48. Shiozaki K, Sato M (2014) Phys Rev B 90:165114. https://doi.org/10.1103/PhysRevB.90.165114
49. Chiu CK, Teo JCY, Schnyder AP, Ryu S (2016) Rev Mod Phys 88:035005. https://doi.org/10.1103/RevModPhys.88.035005
50. Dziawa P, Kowalski BJ, Dybko K, Buczko R, Szczerbakow A, Szot M, Lusakowska E, Balasubramanian T, Wojek BM, Berntsen MH, Tjernberg O, Story T (2012) Nat Mater 11:1023. https://doi.org/10.1038/nmat3449
51. Tanaka Y, Ren Z, Sato T, Nakayama K, Souma S, Takahashi T, Segawa K, Ando Y (2012) Nat Phys 8:800. https://doi.org/10.1038/nphys2442
52. Xu SY, Liu C, Alidoust N, Neupane M, Qian D, Belopolski I, Denlinger J, Wang Y, Lin H, Wray L, Landolt G, Slomski B, Dil J, Marcinkova A, Morosan E, Gibson Q, Sankar R, Chou F, Cava R, Bansil A, Hasan M (2012) Nat Commun 3:1192. https://doi.org/10.1038/ncomms2191
53. Okada Y, Serbyn M, Lin H, Walkup D, Zhou W, Dhital C, Neupane M, Xu S, Wang YJ, Sankar R et al (2013) Science 341(6153):1496
54. Liang T, Gibson Q, Xiong J, Hirschberger M, Koduvayur SP, Cava R, Ong N (2013) arXiv:1307.4022
55. Gyenis A, Drozdov IK, Nadj-Perge S, Jeong OB, Seo J, Pletikosić I, Valla T, Gu GD, Yazdani A (2013) Phys Rev B 88:125414. https://doi.org/10.1103/PhysRevB.88.125414
56. Liu J, Hsieh TH, Wei P, Duan W, Moodera J, Fu L (2014) Nat Mater 13:178. https://doi.org/10.1038/nmat3828
57. Liu J, Duan W, Fu L (2013) Phys Rev B 88:241303. https://doi.org/10.1103/PhysRevB.88.241303
58. Fang C, Gilbert MJ, Xu SY, Bernevig BA, Hasan MZ (2013) Phys Rev B 88:125141. https://doi.org/10.1103/PhysRevB.88.125141
59. Ozawa H, Yamakage A, Sato M, Tanaka Y (2014) Phys Rev B 90:045309. https://doi.org/10.1103/PhysRevB.90.045309

60. Wrasse EO, Schmidt TM (2014) Nano Lett 14(10):5717
61. Liu J, Qian X, Fu L (2015) Nano Lett 15(4):2657
62. Teo JCY, Fu L, Kane CL (2008) Phys Rev B 78:045426. https://doi.org/10.1103/PhysRevB.78.045426
63. Parameswaran SA, Turner AM, Arovas DP, Vishwanath A (2013) Nat Phys 9(5):299
64. Varjas D, de Juan F, Lu YM (2015) Phys Rev B 92:195116. https://doi.org/10.1103/PhysRevB.92.195116
65. Watanabe H, Po HC, Vishwanath A, Zaletel M (2015) Proc Natl Acad Sci USA 112(47):14551
66. Po HC, Watanabe H, Zaletel MP, Vishwanath A (2016) Sci Adv 2(4):e1501782
67. Dong XY, Liu CX (2016) Phys Rev B 93:045429. https://doi.org/10.1103/PhysRevB.93.045429
68. Shiozaki K, Sato M, Gomi K (2016) Phys Rev B 93:195413. https://doi.org/10.1103/PhysRevB.93.195413
69. Liu CX, Zhang RX, VanLeeuwen BK (2014) Phys Rev B 90:085304. https://doi.org/10.1103/PhysRevB.90.085304
70. Fang C, Fu L (2015) Phys Rev B 91:161105(R). https://doi.org/10.1103/PhysRevB.91.161105
71. Shiozaki K, Sato M, Gomi K (2015) Phys Rev B 91:155120. https://doi.org/10.1103/PhysRevB.91.155120
72. Lu L, Fang C, Fu L, Johnson SG, Joannopoulos JD, Soljačić M (2016) Nat Phys 12(4):337
73. Wang Z, Alexandradinata A, Cava RJ, Bernevig BA (2016) Nature 532(7598):189
74. Alexandradinata A, Wang Z, Bernevig BA (2016) Phys Rev X 6:021008. https://doi.org/10.1103/PhysRevX.6.021008
75. Ma J, Yi C, Lv B, Wang Z, Nie S, Wang L, Kong L, Huang Y, Richard P, Zhang P, Yaji K, Kuroda K, Shin S, Weng H, Bernevig BA, Shi Y, Qian T, Ding H (2017) Sci Adv 3(5). https://doi.org/10.1126/sciadv.1602415
76. Ezawa M (2016) Phys Rev B 94:155148. https://doi.org/10.1103/PhysRevB.94.155148
77. Chang PY, Erten O, Coleman P (2017) Nat Phys 13(8):794
78. Zhang D, Shi M, Zhu T, Xing D, Zhang H, Wang J (2019) Phys Rev Lett 122:206401. https://doi.org/10.1103/PhysRevLett.122.206401
79. Li J, Li Y, Du S, Wang Z, Gu BL, Zhang SC, He K, Duan W, Xu Y (2019) Sci Adv 5(6). https://doi.org/10.1126/sciadv.aaw5685
80. Vidal RC, Bentmann H, Peixoto TRF, Zeugner A, Moser S, Min CH, Schatz S, Kißner K, Ünzelmann M, Fornari CI, Vasili HB, Valvidares M, Sakamoto K, Mondal D, Fujii J, Vobornik I, Jung S, Cacho C, Kim TK, Koch RJ, Jozwiak C, Bostwick A, Denlinger JD, Rotenberg E, Buck J, Hoesch M, Diekmann F, Rohlf S, Kalläne M, Rossnagel K, Otrokov MM, Chulkov EV, Ruck M, Isaeva A, Reinert F (2019) Phys Rev B 100:121104. https://doi.org/10.1103/PhysRevB.100.121104
81. Wu J, Liu F, Sasase M, Ienaga K, Obata Y, Yukawa R, Horiba K, Kumigashira H, Okuma S, Inoshita T, Hosono H (2019) Sci Adv 5(11). https://doi.org/10.1126/sciadv.aax9989
82. Li H, Gao SY, Duan SF, Xu YF, Zhu KJ, Tian SJ, Gao JC, Fan WH, Rao ZC, Huang JR, Li JJ, Yan DY, Liu ZT, Liu WL, Huang YB, Li YL, Liu Y, Zhang GB, Zhang P, Kondo T, Shin S, Lei HC, Shi YG, Zhang WT, Weng HM, Qian T, Ding H (2019) Phys Rev X 9:041039. https://doi.org/10.1103/PhysRevX.9.041039
83. Hao YJ, Liu P, Feng Y, Ma XM, Schwier EF, Arita M, Kumar S, Hu C, Lu R, Zeng M, Wang Y, Hao Z, Sun HY, Zhang K, Mei J, Ni N, Wu L, Shimada K, Chen C, Liu Q, Liu C (2019) Phys Rev X 9:041038. https://doi.org/10.1103/PhysRevX.9.041038
84. Chen YJ, Xu LX, Li JH, Li YW, Wang HY, Zhang CF, Li H, Wu Y, Liang AJ, Chen C, Jung SW, Cacho C, Mao YH, Liu S, Wang MX, Guo YF, Xu Y, Liu ZK, Yang LX, Chen YL (2019) Phys Rev X 9:041040. https://doi.org/10.1103/PhysRevX.9.041040
85. Benalcazar WA, Bernevig BA, Hughes TL (2017) Science 357(6346):61
86. Benalcazar WA, Bernevig BA, Hughes TL (2017) Phys Rev B 96:245115. https://doi.org/10.1103/PhysRevB.96.245115
87. Song Z, Fang Z, Fang C (2017) Phys Rev Lett 119:246402. https://doi.org/10.1103/PhysRevLett.119.246402

88. Langbehn J, Peng Y, Trifunovic L, von Oppen F, Brouwer PW (2017) Phys Rev Lett 119:246401. https://doi.org/10.1103/PhysRevLett.119.246401
89. Schindler F, Cook AM, Vergniory MG, Wang Z, Parkin SS, Bernevig BA, Neupert T (2018) Sci Adv 4(6):eaat0346
90. Khalaf E (2018) Phys Rev B 97:205136. https://doi.org/10.1103/PhysRevB.97.205136
91. Khalaf E, Po HC, Vishwanath A, Watanabe H (2018) Phys Rev X 8:031070. https://doi.org/10.1103/PhysRevX.8.031070
92. Matsugatani A, Watanabe H (2018) Phys Rev B 98:205129. https://doi.org/10.1103/PhysRevB.98.205129
93. Ahn J, Yang BJ (2019) Phys Rev B 99:235125. https://doi.org/10.1103/PhysRevB.99.235125
94. Zhang F, Kane CL, Mele EJ (2013) Phys Rev Lett 110:046404. https://doi.org/10.1103/PhysRevLett.110.046404
95. Schindler F, Wang Z, Vergniory MG, Cook AM, Murani A, Sengupta S, Kasumov AY, Deblock R, Jeon S, Drozdov I et al (2018) Nat Phys 14(9):918
96. Xu Y, Song Z, Wang Z, Weng H, Dai X (2019) Phys Rev Lett 122:256402. https://doi.org/10.1103/PhysRevLett.122.256402
97. Wang Z, Wieder BJ, Li J, Yan B, Bernevig BA (2019) Phys Rev Lett 123:186401. https://doi.org/10.1103/PhysRevLett.123.186401
98. Ezawa M (2019) Sci Rep 9(1):5286
99. Yue C, Xu Y, Song Z, Weng H, Lu YM, Fang C, Dai X (2019) Nat Phys 15(6):577
100. Freed DS, Moore GW (2013) Annales Henri Poincaré, vol 14. Springer, pp 1927–2023
101. Read N (2017) Phys Rev B 95:115309. https://doi.org/10.1103/PhysRevB.95.115309
102. Shiozaki K, Sato M, Gomi K (2017) Phys Rev B 95:235425. https://doi.org/10.1103/PhysRevB.95.235425
103. Shiozaki K, Xiong CZ, Gomi K (2018) arXiv:1810.00801
104. Kruthoff J, de Boer J, van Wezel J, Kane CL, Slager RJ (2017) Phys Rev X 7:041069. https://doi.org/10.1103/PhysRevX.7.041069
105. Bradlyn B, Elcoro L, Cano J, Vergniory M, Wang Z, Felser C, Aroyo M, Bernevig BA (2017) Nature 547(7663):298
106. Herring C (1937) Phys Rev 52:365. https://doi.org/10.1103/PhysRev.52.365
107. Murakami S (2007) New J Phys 9(9):356
108. Murakami S, Kuga SI (2008) Phys Rev B 78:165313. https://doi.org/10.1103/PhysRevB.78.165313
109. Burkov AA, Balents L (2011) Phys Rev Lett 107:127205. https://doi.org/10.1103/PhysRevLett.107.127205
110. Wan X, Turner AM, Vishwanath A, Savrasov SY (2011) Phys Rev B 83:205101. https://doi.org/10.1103/PhysRevB.83.205101
111. Volovik GE (2003) The universe in a helium droplet, vol 117. Oxford University Press on Demand
112. Yang KY, Lu YM, Ran Y (2011) Phys Rev B 84:075129. https://doi.org/10.1103/PhysRevB.84.075129
113. Witczak-Krempa W, Kim YB (2012) Phys Rev B 85:045124. https://doi.org/10.1103/PhysRevB.85.045124
114. Okugawa R, Murakami S (2014) Phys Rev B 89:235315. https://doi.org/10.1103/PhysRevB.89.235315
115. Liu J, Vanderbilt D (2014) Phys Rev B 90:155316. https://doi.org/10.1103/PhysRevB.90.155316
116. Soluyanov AA, Gresch D, Wang Z, Wu Q, Troyer M, Dai X, Bernevig BA (2015) Nature 527(7579):495
117. Sun Y, Wu SC, Ali MN, Felser C, Yan B (2015) Phys Rev B 92:161107. https://doi.org/10.1103/PhysRevB.92.161107
118. Hirayama M, Okugawa R, Ishibashi S, Murakami S, Miyake T (2015) Phys Rev Lett 114:206401. https://doi.org/10.1103/PhysRevLett.114.206401

119. Huang SM, Xu SY, Belopolski I, Lee CC, Chang G, Chang TR, Wang B, Alidoust N, Bian G, Neupane M, Sanchez D, Zheng H, Jeng HT, Bansil A, Neupert T, Lin H, Hasan MZ (2016) Proc Natl Acad Sci 113(5):1180. https://doi.org/10.1073/pnas.1514581113
120. Lv BQ, Weng HM, Fu BB, Wang XP, Miao H, Ma J, Richard P, Huang XC, Zhao LX, Chen GF, Fang Z, Dai X, Qian T, Ding H (2015) Phys Rev X 5:031013. https://doi.org/10.1103/PhysRevX.5.031013
121. Xu SY, Belopolski I, Alidoust N, Neupane M, Bian G, Zhang C, Sankar R, Chang G, Yuan Z, Lee CC, Huang SM, Zheng H, Ma J, Sanchez DS, Wang B, Bansil A, Chou F, Shibayev PP, Lin H, Jia S, Hasan MZ (2015) Science 349(6248):613. https://doi.org/10.1126/science.aaa9297
122. Lv B, Xu N, Weng H, Ma J, Richard P, Huang X, Zhao L, Chen G, Matt C, Bisti F et al (2015) Nat Phys 11(9):724
123. Yang L, Liu Z, Sun Y, Peng H, Yang H, Zhang T, Zhou B, Zhang Y, Guo Y, Rahn M et al (2015) Nat Phys 11(9):728
124. Xu SY, Alidoust N, Belopolski I, Yuan Z, Bian G, Chang TR, Zheng H, Strocov VN, Sanchez DS, Chang G et al (2015) Nat Phys 11(9):748
125. Weng H, Fang C, Fang Z, Bernevig BA, Dai X (2015) Phys Rev X 5:011029. https://doi.org/10.1103/PhysRevX.5.011029
126. Huang SM, Xu SY, Belopolski I, Lee CC, Chang G, Wang B, Alidoust N, Bian G, Neupane M, Zhang C et al (2015) Nat Commun 6:7373
127. Xu G, Weng H, Wang Z, Dai X, Fang Z (2011) Phys Rev Lett 107:186806. https://doi.org/10.1103/PhysRevLett.107.186806
128. Wang Z, Vergniory MG, Kushwaha S, Hirschberger M, Chulkov EV, Ernst A, Ong NP, Cava RJ, Bernevig BA (2016) Phys Rev Lett 117:236401. https://doi.org/10.1103/PhysRevLett.117.236401
129. Liu DF, Liang AJ, Liu EK, Xu QN, Li YW, Chen C, Pei D, Shi WJ, Mo SK, Dudin P, Kim T, Cacho C, Li G, Sun Y, Yang LX, Liu ZK, Parkin SSP, Felser C, Chen YL (2019) Science 365(6459):1282. https://doi.org/10.1126/science.aav2873
130. Young SM, Zaheer S, Teo JCY, Kane CL, Mele EJ, Rappe AM (2012) Phys Rev Lett 108:140405. https://doi.org/10.1103/PhysRevLett.108.140405
131. Yang BJ, Nagaosa N (2014) Nat Commun 5:4898
132. Wang Z, Sun Y, Chen XQ, Franchini C, Xu G, Weng H, Dai X, Fang Z (2012) Phys Rev B 85:195320. https://doi.org/10.1103/PhysRevB.85.195320
133. Wang Z, Weng H, Wu Q, Dai X, Fang Z (2013) Phys Rev B 88:125427. https://doi.org/10.1103/PhysRevB.88.125427
134. Liu ZK, Zhou B, Zhang Y, Wang ZJ, Weng HM, Prabhakaran D, Mo SK, Shen ZX, Fang Z, Dai X, Hussain Z, Chen YL (2014) Science 343(6173):864. https://doi.org/10.1126/science.1245085
135. Neupane M, Xu SY, Sankar R, Alidoust N, Bian G, Liu C, Belopolski I, Chang TR, Jeng HT, Lin H et al (2014) Nat Commun 5:3786
136. Borisenko S, Gibson Q, Evtushinsky D, Zabolotnyy V, Büchner B, Cava RJ (2014) Phys Rev Lett 113:027603. https://doi.org/10.1103/PhysRevLett.113.027603
137. Weng H, Liang Y, Xu Q, Yu R, Fang Z, Dai X, Kawazoe Y (2015) Phys Rev B 92:045108. https://doi.org/10.1103/PhysRevB.92.045108
138. Chen Y, Xie Y, Yang SA, Pan H, Zhang F, Cohen ML, Zhang S (2015) Nano Lett 15(10):6974
139. Xie LS, Schoop LM, Seibel EM, Gibson QD, Xie W, Cava RJ (2015) APL Mater 3(8):083602
140. Chan YH, Chiu CK, Chou MY, Schnyder AP (2016) Phys Rev B 93:205132. https://doi.org/10.1103/PhysRevB.93.205132
141. Kim Y, Wieder BJ, Kane CL, Rappe AM (2015) Phys Rev Lett 115:036806. https://doi.org/10.1103/PhysRevLett.115.036806
142. Yu R, Weng H, Fang Z, Dai X, Hu X (2015) Phys Rev Lett 115:036807. https://doi.org/10.1103/PhysRevLett.115.036807
143. Schoop LM, Ali MN, Straßer C, Topp A, Varykhalov A, Marchenko D, Duppel V, Parkin SS, Lotsch BV, Ast CR (2016) Nat Commun 7:11696

144. Neupane M, Belopolski I, Hosen MM, Sanchez DS, Sankar R, Szlawska M, Xu SY, Dimitri K, Dhakal N, Maldonado P, Oppeneer PM, Kaczorowski D, Chou F, Hasan MZ, Durakiewicz T (2016) Phys Rev B 93:201104. https://doi.org/10.1103/PhysRevB.93.201104
145. Hu J, Tang Z, Liu J, Liu X, Zhu Y, Graf D, Myhro K, Tran S, Lau CN, Wei J, Mao Z (2016) Phys Rev Lett 117:016602. https://doi.org/10.1103/PhysRevLett.117.016602
146. Huang H, Liu J, Vanderbilt D, Duan W (2016) Phys Rev B 93:201114. https://doi.org/10.1103/PhysRevB.93.201114
147. Li R, Ma H, Cheng X, Wang S, Li D, Zhang Z, Li Y, Chen XQ (2016) Phys Rev Lett 117:096401. https://doi.org/10.1103/PhysRevLett.117.096401
148. Hirayama M, Okugawa R, Miyake T, Murakami S (2017) Nat Commun 8:14022
149. Haldane FDM, Raghu S (2008) Phys Rev Lett 100:013904. https://doi.org/10.1103/PhysRevLett.100.013904
150. Raghu S, Haldane FDM (2008) Phys Rev A 78:033834. https://doi.org/10.1103/PhysRevA.78.033834
151. Wang Z, Chong YD, Joannopoulos JD, Soljačić M (2008) Phys Rev Lett 100:013905. https://doi.org/10.1103/PhysRevLett.100.013905
152. Wang Z, Chong Y, Joannopoulos JD, Soljačić M (2009) Nature 461(7265):772
153. Hafezi M, Demler EA, Lukin MD, Taylor JM (2011) Nat Phys 7(11):907
154. Khanikaev AB, Mousavi SH, Tse WK, Kargarian M, MacDonald AH, Shvets G (2013) Nat Mater 12(3):233
155. Hafezi M, Mittal S, Fan J, Migdall A, Taylor J (2013) Nat Photonics 7(12):1001
156. Mittal S, Fan J, Faez S, Migdall A, Taylor JM, Hafezi M (2014) Phys Rev Lett 113:087403. https://doi.org/10.1103/PhysRevLett.113.087403
157. Ma T, Khanikaev AB, Mousavi SH, Shvets G (2015) Phys Rev Lett 114:127401. https://doi.org/10.1103/PhysRevLett.114.127401
158. Mittal S, Ganeshan S, Fan J, Vaezi A, Hafezi M (2016) Nat Photonics 10(3):180
159. Slobozhanyuk A, Mousavi SH, Ni X, Smirnova D, Kivshar YS, Khanikaev AB (2017) Nat Photonics 11(2):130
160. Yang Y, Gao Z, Xue H, Zhang L, He M, Yang Z, Singh R, Chong Y, Zhang B, Chen H (2019) Nature 565(7741):622
161. Lu L, Fu L, Joannopoulos JD, Soljačić M (2013) Nat Photonics 7(4):294
162. Ochiai T (2017) Phys Rev A 96:043842. https://doi.org/10.1103/PhysRevA.96.043842
163. Lu L, Gao H, Wang Z (2018) Nat Commun 9(1):1
164. Onose Y, Ideue T, Katsura H, Shiomi Y, Nagaosa N, Tokura Y (2010) Science 329(5989):297. https://doi.org/10.1126/science.1188260
165. Mook A, Henk J, Mertig I (2014) Phys Rev B 89:134409. https://doi.org/10.1103/PhysRevB.89.134409
166. Lee H, Han JH, Lee PA (2015) Phys Rev B 91:125413. https://doi.org/10.1103/PhysRevB.91.125413
167. Hirschberger M, Chisnell R, Lee YS, Ong NP (2015) Phys Rev Lett 115:106603. https://doi.org/10.1103/PhysRevLett.115.106603
168. Chisnell R, Helton JS, Freedman DE, Singh DK, Bewley RI, Nocera DG, Lee YS (2015) Phys Rev Lett 115:147201. https://doi.org/10.1103/PhysRevLett.115.147201
169. Nakata K, Kim SK, Klinovaja J, Loss D (2017) Phys Rev B 96:224414. https://doi.org/10.1103/PhysRevB.96.224414
170. Li FY, Li YD, Kim YB, Balents L, Yu Y, Chen G (2016) Nat Commun 7:12691
171. Mook A, Henk J, Mertig I (2016) Phys Rev Lett 117:157204. https://doi.org/10.1103/PhysRevLett.117.157204
172. Mook A, Henk J, Mertig I (2017) Phys Rev B 95:014418. https://doi.org/10.1103/PhysRevB.95.014418
173. Li K, Li C, Hu J, Li Y, Fang C (2017) Phys Rev Lett 119:247202. https://doi.org/10.1103/PhysRevLett.119.247202
174. Yao W, Li C, Wang L, Xue S, Dan Y, Iida K, Kamazawa K, Li K, Fang C, Li Y (2018) Nat Phys 14(10):1011

175. Zhang L, Ren J, Wang JS, Li B (2010) Phys Rev Lett 105:225901. https://doi.org/10.1103/PhysRevLett.105.225901
176. Kane C, Lubensky T (2014) Nat Phys 10(1):39
177. Yang Z, Gao F, Shi X, Lin X, Gao Z, Chong Y, Zhang B (2015) Phys Rev Lett 114:114301. https://doi.org/10.1103/PhysRevLett.114.114301
178. Wang P, Lu L, Bertoldi K (2015) Phys Rev Lett 115:104302. https://doi.org/10.1103/PhysRevLett.115.104302
179. Xiao M, Chen WJ, He WY, Chan CT (2015) Nat Phys 11(11):920
180. Süsstrunk R, Huber SD (2015) Science 349(6243):47
181. Mousavi SH, Khanikaev AB, Wang Z (2015) Nat Commun 6:8682
182. Nash LM, Kleckner D, Read A, Vitelli V, Turner AM, Irvine WT (2015) Proc Natl Acad Sci 112(47):14495
183. Huber SD (2016) Nat Phys 12(7):621
184. Rocklin DZ, Chen BG, Falk M, Vitelli V, Lubensky TC (2016) Phys Rev Lett 116:135503. https://doi.org/10.1103/PhysRevLett.116.135503
185. Süsstrunk R, Huber SD (2016) Proc Natl Acad Sci 113(33):E4767
186. Liu Y, Xu Y, Zhang SC, Duan W (2017) Phys Rev B 96:064106. https://doi.org/10.1103/PhysRevB.96.064106
187. Zhang T, Song Z, Alexandradinata A, Weng H, Fang C, Lu L, Fang Z (2018) Phys Rev Lett 120:016401. https://doi.org/10.1103/PhysRevLett.120.016401
188. Weick G, Woollacott C, Barnes WL, Hess O, Mariani E (2013) Phys Rev Lett 110:106801. https://doi.org/10.1103/PhysRevLett.110.106801
189. Jin D, Lu L, Wang Z, Fang C, Joannopoulos JD, Soljačić M, Fu L, Fang NX (2016) Nat Commun 7:13486
190. Jin D, Christensen T, Soljačić M, Fang NX, Lu L, Zhang X (2017) Phys Rev Lett 118:245301. https://doi.org/10.1103/PhysRevLett.118.245301

# Chapter 3
# Weyl Semimetals and Spinless $Z_2$ Magnetic Topological Crystalline Insulators with Glide Symmetry

In this chapter, we study phase transitions between a spinless topological crystalline insulator (TCI) and a normal insulator (NI) phases, upon changing a parameter in the magnetic system. We assume that the glide symmetry is preserved in the phase transition. First of all, we construct a theory describing such a phase transition based on an effective model. We find that the TCI-NI phase transition is always intervened by a spinless Weyl semimetal (WSM) phase in general, which is a semimetal phase with nondegenerate Dirac cones at or around the Fermi energy. In general, all the bands in spinless systems are free from degeneracy in this case, and a gap closing between a single conduction band and a single valence band gives rise to a pair creation of Weyl nodes in general; thus the WSM phase naturally appears. Since the Weyl nodes (the apices of the Dirac cones) are topological, i.e., either monopoles or antimonopoles in momentum space, the Weyl nodes survive within a finite range of the parameter $m$, until they are pairwise annihilated. We relate the trajectory of the Weyl nodes within the WSM phase, and the change of the $Z_2$ topological invariant for glide-symmetric magnetic systems. Then we support this conclusion by a calculation on a lattice model. The results totally agree with our theory of topological phase transition in spinless magnetic glide-symmetric systems.

## 3.1 General Theory: Phase Transition for the Glide-Symmetric Magnetic Systems

In the present section, we study basic properties of the glide-symmetric magnetic TCI, and then we construct a general theory of the TCI-NI phase transition to show that the WSM phase necessarily appears between the two topologically distinct phases in the spinless system. We consider a general spinless system with glide symmetry in the absence of time-reversal symmetry, with a parameter $m$ which controls the



H. Kim, *Glide-Symmetric $Z_2$ Magnetic Topological Crystalline Insulators*,
Springer Theses, https://doi.org/10.1007/978-981-16-9077-8_3

TCI-NI topological phase transition. When changing the parameter $m$, we assume that glide symmetry is preserved. This parameter $m$ can be physically any parameter in the system, such as atomic compositions or pressure. In order to study the TCI-NI phase transition, we first start with the system with a bulk gap and assume that the gap is closed by changing $m$. We then discuss in which cases the system will open a gap again by the change of $m$. In thi end, this shows that a spinless WSM phase should come out between the TCI and NI phases in glide-symmetric systems. Our interest here is how the evolution of the Weyl nodes are related to the $Z_2$ topological invariant characterizing the magnetic TCI phase enriched by glide symmetry.

### 3.1.1 $Z_2$ Topological Invariant for Glide-Symmetric Magnetic Systems

Here we briefly study the magnetic TCI enriched by glide symmetry proposed in Refs. [1, 2]. To be concrete, let us start with a three-dimensional (3D) system invariant under a glide operation

$$\hat{G}_y = \left\{ M_y \middle| \frac{1}{2}\hat{\mathbf{z}} \right\}, \tag{3.1}$$

where $M_y$ represents the mirror reflection with respect to the $xz$ plane, and $\hat{\mathbf{z}}$ is a unit vector along the $z$ axis. Henceforth, we set the lengths of the primitive vectors to be unity. Then the Bloch Hamiltonian $H(\mathbf{k})$ for the glide-symmetric system satisfies the equation

$$G_y(k_z)H(k_x, k_y, k_z)G_y(k_z)^{-1} = H(k_x, -k_y, k_z), \tag{3.2}$$

where $G_y(k_z)$ is the $k$-dependent glide operator representing $\hat{G}_y$. Because the Hamiltonian commutes with the glide operator on glide-invariant planes $k_y = 0$ and $k_y = \pi$, the Hamiltonian can be block-diagonalized into two blocks which are featured by the eigenvalues of $G_y(k_z)$. As we have $\hat{G}_y^2 = \{E|\hat{\mathbf{z}}\}$, the eigenvalues of $G_y(k_z)$ in spinless systems are given by

$$g_\pm(k_z) = \pm e^{-ik_z/2}. \tag{3.3}$$

Thus, the two branches for those eigenvalues are related to each other,

$$g_\pm(k_z \pm 2\pi) = g_\mp(k_z), \tag{3.4}$$

and are interchanged when an eigenstate goes across the *branch cut* because the eigenvalues have $4\pi$ periodicity for $k_z$. This is a remarkable property of systems with nonsymmorphic symmetries with a fractional translation. Due to this $4\pi$ periodicity, the minimum number of bands in the conduction band is two, and that for the valence band is also two. Thus the minimal number of bands to show the TCI-NI phase

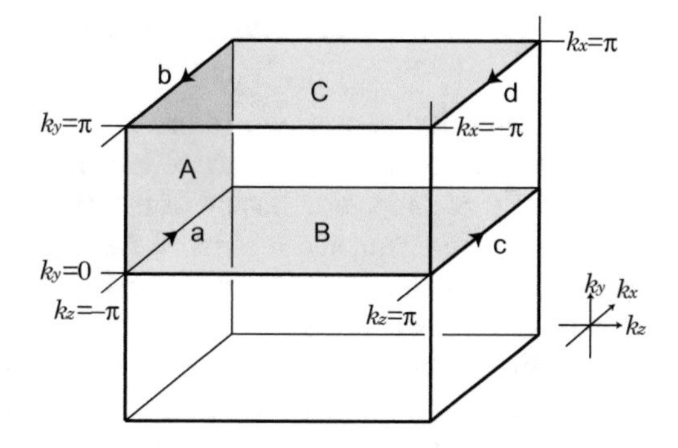


**Fig. 3.1** Three-dimensional Brillouin zone. The Berry curvatures are computed in the regions A, B, and C, and the Berry phases are computed along the arrows a and b in Eq. (3.5)

transition is four, and we consider $4 \times 4$ Hamiltonian $H(\mathbf{k}, m)$, where $m$ is the parameter to drive the TCI-NI phase transition.

One can define the $Z_2$ topological invariant for such glide-symmetric systems in the absence of time-reversal symmetry. Although the integral of Berry curvature for each glide sector on the glide-invariant planes can be defined, this quantity is not a quantized topological invariant because of the existence of the branch cut. Instead, a 3D gapped glide-symmetric magnetic system is characterized by the $Z_2$ topological invariant [1, 2] defined as

$$\begin{aligned}\nu = \frac{1}{2\pi}\Bigg[&-2\int_{-\pi}^{\pi} dk_x \mathrm{tr}\mathcal{A}_x^+(k_x,0,-\pi) + 2\int_{-\pi}^{\pi} dk_x \mathrm{tr}\mathcal{A}_x^+(k_x,\pi,-\pi) \\ &+\int_{-\pi}^{\pi} dk_x \int_{-\pi}^{\pi} dk_z \mathrm{tr}\mathcal{F}_{zx}^-(k_x,0,k_z) - \int_{-\pi}^{\pi} dk_x \int_{-\pi}^{\pi} dk_z \mathrm{tr}\mathcal{F}_{zx}^-(k_x,\pi,k_z) \\ &+\int_{0}^{\pi} dk_y \int_{-\pi}^{\pi} dk_x \mathrm{tr}\mathcal{F}_{xy}(k_x,k_y,-\pi)\Bigg] \pmod 2. \end{aligned} \tag{3.5}$$

The integral regions shown in Fig. 3.1. We have defined the Berry connections

$$\mathcal{A}_{i,mn}(\mathbf{k}) \equiv i\langle u_{m\mathbf{k}}|\partial_{k_i}|u_{n\mathbf{k}}\rangle, \tag{3.6}$$
$$\mathcal{A}_{i,mn}^{\pm}(\mathbf{k}) \equiv i\langle u_{m\mathbf{k}}^{\pm}|\partial_{k_i}|u_{n\mathbf{k}}^{\pm}\rangle, \tag{3.7}$$

and the corresponding Berry curvatures

$$\mathcal{F}_{ij}(\mathbf{k}) = \partial_{k_i}\mathcal{A}_j - \partial_{k_j}\mathcal{A}_i + i\left[\mathcal{A}_i, \mathcal{A}_j\right], \tag{3.8}$$
$$\mathcal{F}_{ij}^{\pm}(\mathbf{k}) = \partial_{k_i}\mathcal{A}_j^{\pm} - \partial_{k_j}\mathcal{A}_i^{\pm} + i\left[\mathcal{A}_i^{\pm}, \mathcal{A}_j^{\pm}\right], \tag{3.9}$$

where $m, n$ are the band indices, $|u_{n\mathbf{k}}\rangle$ is the Bloch wavefunction, and $|u_{n\mathbf{k}}^{\pm}\rangle$ is the Bloch wavefunction belonging to the glide sector with the glide eigenvalue $g_{\pm}(k_z) = \pm e^{-ik_z/2}$.

### 3.1.2 TCI-NI Phase Transition and Weyl Semimetal

We next consider the case where the TCI-NI phase transition occurs by changing a parameter $m$ controlling the phase transition. Since the TCI-NI phase transition neccessarily accompanies closing of the gap, suppose we start with the NI phase, and change the parameter $m$ to close the gap. Because all the states are generally free from degeneracies, we need to consider a single conduction band and a single valence band. Such a closing of the band gap should be in general a pair creation of the Weyl nodes [3]. Once the Weyl nodes are created, the Weyl nodes migrate in momentum space due to their topological stability. This phase is generally a Weyl semimetal phase. The Weyl nodes are at the same energy if all the Weyl nodes are related to each other by symmetry operations; otherwise, not all the Weyl nodes are at the same energy. After a further change of $m$, if all the Weyl nodes disappear by pair annihilation, the system goes into a bulk insulator again. This bulk insulating phase may be the spinless glide-symmetric TCI phase, or the NI phase. Because the Weyl nodes carry topological charges, which phase is realized relies on the trajectory of the Weyl nodes within the WSM phase.

The glide symmetry restricts the motion of the Weyl nodes. We note that the eigenstates are classified into the two glide sectors *only on the glide plane*. Therefore, a pair creation on the glide plane is classified into two cases, Fig. 3.2a, c, in terms of the glide sector. Here, $H_{\pm}$ means the pair creation/annihilation on the glide sector with the glide eigenvalues of $g_{\pm}(k_z) = \pm e^{-ik_z/2}$. In the same way, a pair annihilation on the glide plane is classified into two cases, Fig. 3.2b, d. By combining (a)/(c) with (b)/(d), there are several patterns for annihilating all the Weyl nodes; thereby, the system turns into a bulk-insulating phase again. For instance, suppose Weyl nodes are created in the $H_+$ sector. After the change of the parameter $m$, we can consider two cases of pair annihilation: Weyl nodes are annihilated at the $H_+$ sector (Fig. 3.2e) or the $H_-$ sector (Fig. 3.2f). From the third or forth term of the glide-$Z_2$ invariant in Eq. (3.5), the latter case (Fig. 3.2f) is accompanied by the change of $\nu$ by one, i.e., the TCI-NI phase transition, while the former case (Fig. 3.2e) is not. Intuitively, one can understand this since the trajectory of the Weyl nodes can be continuously deformed to null in the former case, while that is topologically nontrivial in the latter case. We exclude the cases with Weyl nodes coming exactly on the branch cut, since it does not happen unless other parameters are finely tuned.

Moreover, we should mention the existence of branch cut, because the glide sectors are switched when we go across the branch cut, from the nature of the glide eigenvalues $g_{\pm}(k_z + 2\pi) = g_{\pm}(k_z - 2\pi) = g_{\mp}(k_z)$. Due to this switching, when the pair creation and pair annihilation occur at the same $H_+$ sector, but a branch cut exists between the positions of the pair creation and pair annihilation (Fig. 3.2g), the glide-$Z_2$ invariant is changed between the two bulk-insulating phases, because it is equivalent to Fig. 3.2f. In this case, the last term in the formula of the glide-$Z_2$ invariant (3.5) is changed by one.

In addition to these cases, there are numerous cases for the evolution of Weyl nodes. From Eq. (3.5), the change of the glide-$Z_2$ invariant $\Delta\nu$ is obtained from the

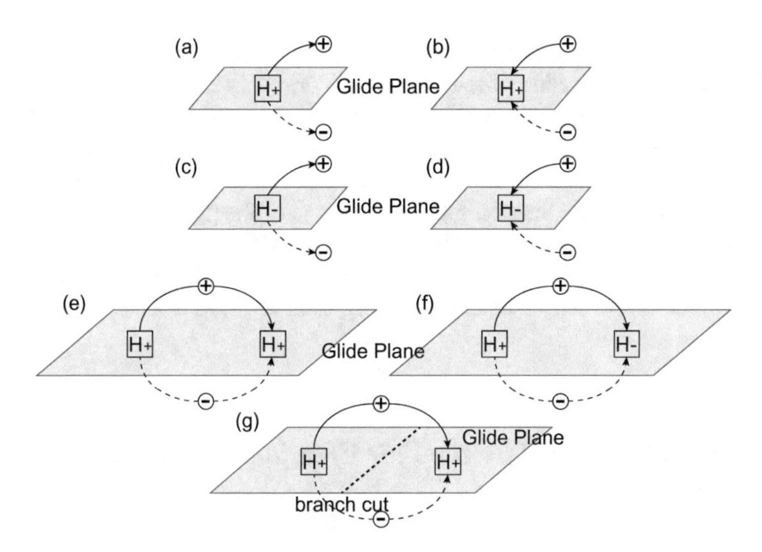


**Fig. 3.2** **a–d** Possible cases of pair creation/annihilation of Weyl nodes. The directions of the arrows illustrate the trajectories of the Weyl nodes with the change of the parameter $m$. $H_+$ ($H_-$) indicates the subspace sector with the glide eigenvalue $g_\pm(k_z) = \pm e^{-ik_z/2}$. The signs $+$ $(-)$ in the circles denote the monopole (antimonopole). **e–g** Three examples of the motions of the Weyl nodes within the Weyl semimetal phase, in the transition between two bulk-insulating phases. Because of the existence of the glide symmetry, their trajectories are symmetric with respect to the glide plane. The trajectories of Weyl nodes are topologically trivial in **e**, and nontrivial in **f** and **g**; it means that the TCI-NI phase transition occurs in **f** and **g** but not in **e**

trajectory of the Weyl nodes in the following way. First, according to Ref. [4], we flip the orientation of the trajectory of the antimonopole, whereas that of the monopole is left unchanged; this always leads us to a single oriented loop [4]. Then the change of the glide-$Z_2$ invariant $\Delta\nu$ is generally written as

$$\Delta\nu = N_{\mathrm{A}} + N_{\mathrm{B}}^{-} + N_{\mathrm{C}}^{-} \pmod 2, \tag{3.10}$$

where $N_{\mathrm{A}}$ is the number of crossings between the loop and the $k_z = -\pi$ plane, and $N_{\mathrm{B}}^{-}$ ($N_{\mathrm{C}}^{-}$) is the number of crossings between the loop and the $k_y = 0$ ($k_y = \pi$) plane within the $H_-$ sector. Here we note that we do not need to consider contributions in Eq. (3.10) from the first and second terms in Eq. (3.5). The reason is as follows. When nodal points (i.e., Weyl nodes) or nodal lines go across the paths a or b in Fig. 3.1 by changing of the parameter $m$, these two terms change. Nodal points or lines here refer to a set of **k** points where the gap is closed. Nonetheless, the nodal points in general do not go across the paths a or b unless other parameters are finely tuned. Moreover, nodal lines at generic **k** points do not appear in glide-symmetric TCIs, because of the broken time-reversal symmetry.

Here we comment on time-reversal symmetry in such spinless glide-symmetric systems. As mentioned in Ref. [1], the glide-$Z_2$ invariant is trivial when the time-reversal symmetry is preserved. It can be seen from our viewpoint of trajectories of Weyl nodes. Because the time-reversal operator $\mathcal{T} \equiv K$ (complex conjugation) satisfies $\mathcal{T}G_y = G_y^{-1}\mathcal{T}$, when a pair creation of Weyl nodes occurs within the glide sector $H_+$ at $\mathbf{k}_0$, there simultaneously occurs a pair creation within the same glide sector $H_+$ at $-\mathbf{k}_0$. It is also the case for the $H_-$ sector. Hence the quantities $N_{\mathrm{B}}^-$ and $N_{\mathrm{C}}^-$ are always even, implying that the change of the glide-$Z_2$ invariant is zero. Namely, in time-reversal invariant spinless systems with glide symmetry, one cannot have a NI-WSM-TCI phase transition, in accordance with the argument in Ref. [1]. We encounter several cases in the following section using a simple tight-binding model.

### 3.1.3 *Effective Model for the Pair Creation/Annihilation of Weyl Nodes*

We can construct an effective model describing a pair creation/annihilation of Weyl nodes on the glide plane. On the glide plane the Hamiltonian $H(\mathbf{k}, m)$ satisfies

$$G_y H(k_x, k_y, k_z, m) G_y^{-1} = H(k_x, -k_y, k_z, m). \tag{3.11}$$

As an example, we focus on the pair creation/annihilation on the glide plane $k_y = 0$, i.e., at $\mathbf{k}_0 = (k_{0x}, 0, k_{0z})$, $m = m_0$. For simplicity, we assume there are no symmetries except for the glide symmetry. All the states are then free from degeneracies, and the gap closes between a single conduction band and a single valence band. Hence we need to consider a $2 \times 2$ effective Hamiltonian:

$$H(\mathbf{k}, m) = \epsilon(\mathbf{k}, m) + \begin{pmatrix} a(\mathbf{k}, m) & b(\mathbf{k}, m) \\ b^*(\mathbf{k}, m) & -a(\mathbf{k}, m) \end{pmatrix}, \tag{3.12}$$

where $\epsilon(\mathbf{k}, m)$ and $a(\mathbf{k}, m)$ are real functions and $b(\mathbf{k}, m)$ is a complex function. We then expand $H(\mathbf{k}, m)$ in terms of $\delta\mathbf{k} \equiv \mathbf{k} - \mathbf{k}_0$ and $\delta m = m - m_0$. At $\mathbf{k}_0, m_0$, the Hamiltonian has degenerate eigenstates, and so the entities of the matrix have no zeroth order terms. We note that at the gap closing at $\mathbf{k}_0, m_0$, the two states have the same glide eigenvalue. Thus the matrix representation of $\hat{G}_y$ is proportional to the identity matrix. From Eq. (3.11), $a$ and $b$ are then even functions of $k_y = \delta k_y$ and we can write

$$a(\mathbf{k}, m) = a_1 \delta k_x + a_2 \delta k_y^2 + a_3 \delta k_z + a_0 \delta m, \tag{3.13}$$

$$b(\mathbf{k}, m) = b_1 \delta k_x + b_2 \delta k_y^2 + b_3 \delta k_z + b_0 \delta m, \tag{3.14}$$

up to the linear order in $\delta k_x, \delta k_z$, and $m$, and the quadratic order in $\delta k_y$. The gap closes when $a = 0$, $\mathrm{Re}b = 0$, and $\mathrm{Im}b = 0$ are satisfied simultaneously. Namely,

$$\begin{pmatrix} a_1 & a_3 & a_0 \\ \mathrm{Re}b_1 & \mathrm{Re}b_3 & \mathrm{Re}b_0 \\ \mathrm{Im}b_1 & \mathrm{Im}b_3 & \mathrm{Im}b_0 \end{pmatrix} \begin{pmatrix} \delta k_x \\ \delta k_z \\ \delta m \end{pmatrix} = \begin{pmatrix} -a_2 \\ -\mathrm{Re}b_2 \\ -\mathrm{Im}b_2 \end{pmatrix} \delta k_y^2. \tag{3.15}$$

This equation yields a solution with the form $\delta k_x = A_x \delta k_y^2, \delta k_z = A_z \delta k_y^2, \delta m = A_0 \delta k_y^2$, where $A_x$, $A_z$, and $A_0$ are real constants. As $m$ is increased, this effective model describes a pair creation ($A_0 > 0$) or a pair annihilation ($A_0 < 0$) of Weyl nodes on the glide plane. For instance, when $A_0 > 0$, a pair of Weyl nodes is created at $(\frac{A_x}{A_0}\delta m, \pm\sqrt{\frac{\delta m}{A_0}}, \frac{A_z}{A_0}\delta m)$ only when $\delta m > 0$. From the relations $\delta k_x = A_x \delta k_y^2$ and $\delta k_z = A_z \delta k_y^2$, the trajectory of the Weyl nodes is parabolic around $\mathbf{k}_0$, $m_0$ and is symmetric with respect to the glide plane.

## 3.2 Model Calculation: 3D Fang-Fu Lattice Model with an Additional Term

In the present section, we consider the 3D tight-binding model proposed by Fang and Fu. In order to describe general properties of the NI-WSM-TCI phase transition, we add a term to the model to lift spurious double degeneracy of the original model. The results confirm the general arguments in the previous section.

### 3.2.1 *Model*

To demonstrate our scenario, we use the tight-binding model proposed by Fang and Fu [1] with an additional term. Our model is a four-band spinless system on a 3D orthorhombic lattice with two sublattices and two orbitals, written as the following Hamiltonian:

$$\begin{aligned} H(\mathbf{k}) &= (m - t_0 \cos k_x - t_0' \cos k_y - t_0'' \cos k_z)\sigma_z \\ &\quad + t'' \sin \frac{k_z - \phi}{2} \left( \cos \frac{k_z}{2} \sigma_x \tau_x + \sin \frac{k_z}{2} \sigma_x \tau_y \right) \\ &\quad + t \sin k_x \sigma_y + t' \sin k_y \sigma_x \tau_z + p \sigma_x, \end{aligned} \tag{3.16}$$

where $\boldsymbol{\sigma}$ and $\boldsymbol{\tau}$ are Pauli matrices describing the orbital and lattice degrees of freedom, respectively, and the coefficients $t_0$, $t_0'$, $t_0''$, $t$, $t'$, $t''$, and $p$ are constants. It is invariant under the glide operator given by

$$G_y(k_z) = e^{-ik_z/2} \left( \cos \frac{k_z}{2} \tau_x + \sin \frac{k_z}{2} \tau_y \right), \tag{3.17}$$

which satisfies

$$G_y(k_z)H(k_x,k_y,k_z)G_y^{-1}(k_z) = H(k_x,-k_y,k_z), \tag{3.18}$$

$$G_y^2(k_z) = e^{-ik_z}. \tag{3.19}$$

We assume $t_0$, $t_0'$, $t_0''$, $t$, $t'$, and $t''$ to be positive for simplicity.

In the absence of the last term ($p = 0$), it is exactly the model proposed by Fang and Fu [1], in which all the states are doubly degenerate. Nonetheless, this double degeneracy for every **k** does not originate from crystallographic symmetries, and therefore we cannot expect this degeneracy in general spinless glide-symmetric systems. Here we briefly mention the symmetry of the Fang-Fu model, causing the double degeneracy. We define the following operations:

$$O = c(t\sigma_y + M\sigma_z)\left(t'\cos\frac{k_z}{2}\tau_x + t'\sin\frac{k_z}{2}\tau_y - t''\tau_z\right), \tag{3.20}$$

$$O' = c(t\sigma_y + M\sigma_z)\left(t''\cos\frac{k_z}{2}\tau_x + t''\sin\frac{k_z}{2}\tau_y + t'\tau_z\right), \tag{3.21}$$

where $c = \frac{1}{\sqrt{t^2+M^2}}\frac{1}{\sqrt{t'^2+t''^2}}$. It satisfies

$$[O,H] = 0, \quad [O',H] = 0, \quad \{O,O'\} = 0, \quad O^2 = 1 = O'^2. \tag{3.22}$$

The eigenstates of the Hamiltonian thus can be chosen as eigenstates of $O$. The eigenvalues of $O$ are either $+1$ or $-1$, and $O'$ switches the signs of the eigenvalues of $O$ without changing the energy. Hence, all the eigenstates are doubly degenerate, having the opposite signs of the eigenvalues of $O$. This symmetry described by $O'$ and $O$ cannot come from any crystallographic symmetries, and one cannot expect such double degeneracy in crystals. Thus we added the term $p\tau_x$ to remove the degeneracy. Our question is how the phase transition between the NI and the TCI phases occurs in glide-symmetric systems.

The energy eigenvalues of the Hamiltonian (3.16) are

$$E(\mathbf{k}) = \pm\sqrt{f_0^2 + t^2\sin^2 k_x + f^2}, \tag{3.23}$$

where $f_0 = m - t_0\cos k_x - t_0'\cos k_y - t_0''\cos k_z$ and $f = \sqrt{t'^2\sin^2 k_y + t''^2\sin^2\frac{k_z-\phi}{2}} \pm p$. We note that the energy spectrum is symmetric with respect to $E = 0$. We set $E_F = 0$ and focus on the bulk band gap at $E = 0$. Particularly, when $p = 0$ [1], all the bands are doubly degenerate and the gap closes at

$$k_z = \phi, \ (k_x,k_y) = (0,0), (0,\pi), (\pi,0), (\pi,\pi), \tag{3.24}$$

$$m = t_0\cos k_x + t_0'\cos k_y + t_0''\cos\phi, \tag{3.25}$$

as described in Ref. [1]. In this case, the bulk gap closes at four values of $m$, $m = \pm t_0 \pm t_0' + t_0'' \cos\phi$, accompanied by NI-TCI phase transitions. Among them, we focus on the gap closing at $m = t_0 + t_0' + t_0'' \cos\phi$ and $\mathbf{k} = (0, 0, \phi)$, as an example, and we add the additional $p$ term to see emergence of the WSM phase.

Artificial double degeneracy in the Fang-Fu model is eliminated by adding the $p$ term. We below show that a spinless WSM phase intervenes between the TCI and the NI phases. When $p \neq 0$, the band gap closes where

$$t'^2 \sin^2 k_y + t''^2 \sin^2 \frac{k_z - \phi}{2} = p^2,\ k_x = 0, \pi, \tag{3.26}$$

$$m = t_0 \cos k_x + t_0' \cos k_y + t_0'' \cos k_z. \tag{3.27}$$

Hence the closing of the gap at $\mathbf{k} = (0, 0, \phi)$ and $m = t_0 + t_0' + t_0'' \cos\phi$ in the case of $p = 0$ is split because of the lifting of the degeneracy by the additional $p$ term. Now the gap is closed within a finite range of the value of $m$. As can be shown explicitly, the gap-closing points are Weyl nodes within this region, and therefore the system is in the WSM phase in this region of $m$. For instance, we show the phase diagram near $m = 3$, $p = 0$ in the $m$-$p$ plane in Fig. 3.3a for $t_0 = t_0' = t_0'' = t = t' = t'' = 1$. Unless $p = 0$, we can see that the WSM phase extends over a finite range of $m$. We show how this phase diagram results in the following.

### 3.2.2 *Trajectories of Weyl Nodes*

Here we illustrate behaviors of the Weyl nodes in the WSM phase of our model. First, we note that the curve represented by Eq. (3.26) describes a trajectory of the Weyl nodes in momentum space. The trajectory of the Weyl nodes is symmetric with respect to the glide plane, and the monopoles and antimonopoles are switched by glide operation. This curve crosses the glide plane $k_y = 0$ at two points $\mathbf{k}^{\pm} = (0, 0, \phi \mp 2 \arcsin \frac{p}{t''})$, when $m^{\pm} = t_0 + t_0' + t_0'' \cos k_z^{\pm}$. For simplicity, we here assume $|p/t''| < 1$. In this parameter region, the Hamiltonian reads

$$H(\mathbf{k}^{\pm}, m^{\pm}) = \mp p \left( \cos \frac{k_z^{\pm}}{2} \sigma_x + \sin \frac{k_z^{\pm}}{2} \sigma_y \right) \tau_x + p \tau_x, \tag{3.28}$$

and the gap closes at $E = 0$ with the eigenstates

$$\Psi_{\pm}^{(1)} = \frac{1}{2}^t (1, 1, \pm e^{ik_z^{\pm}/2}, \pm e^{ik_z^{\pm}/2}), \tag{3.29}$$

$$\Psi_{\pm}^{(2)} = \frac{1}{2}^t (1, -1, \pm e^{ik_z^{\pm}/2}, \mp e^{ik_z^{\pm}/2}), \tag{3.30}$$

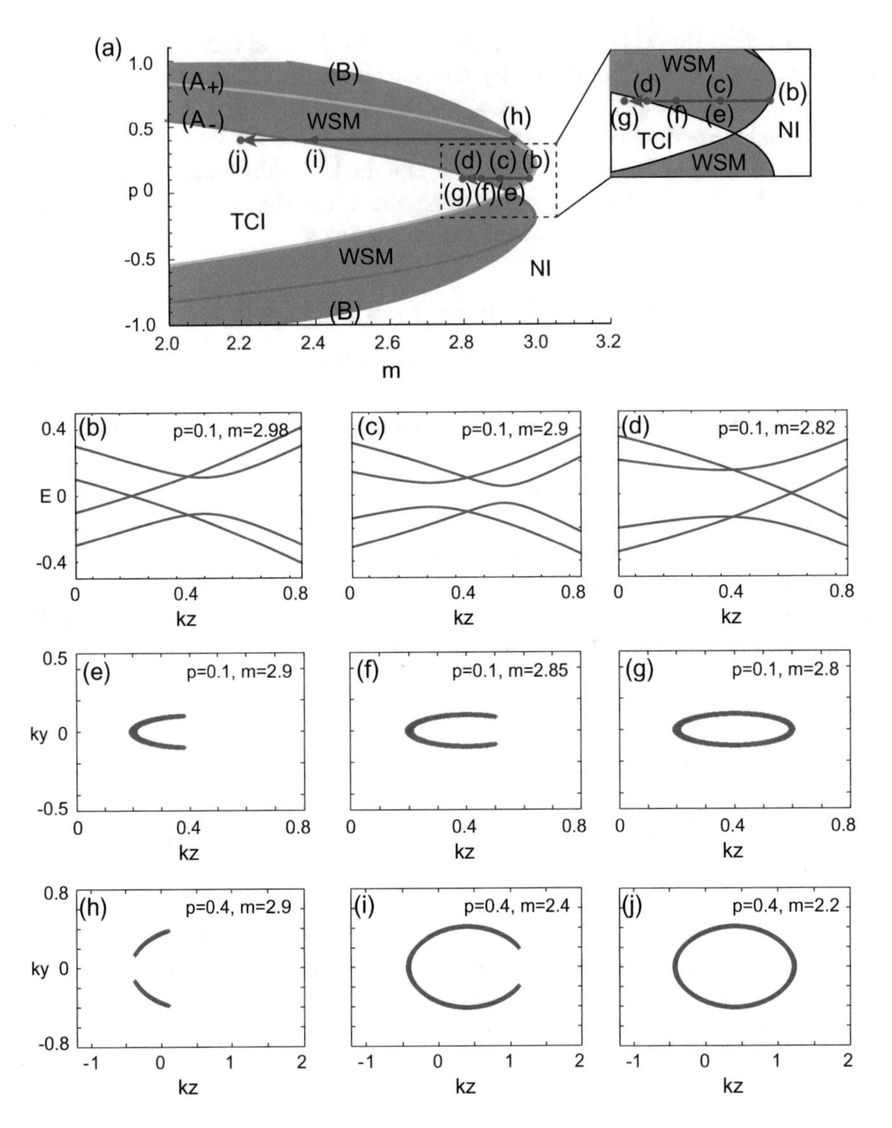


**Fig. 3.3** **a** Phase diagram of the present model in the $m$-$p$ plane with $t_0 = t_0' = t_0'' = t = t' = t'' = 1$ and $\phi = 0.4$. The bulk energy dispersions become gapless on the lines ($A_\pm$) and (B). On the line ($A_\pm$), the bulk bands become gapless on the glide plane within the $H_\pm$ sector. On the line (B), the bulk band gap closes outside of the glide plane. **b–d** Bulk band structures on the $k_x = k_y = 0$ line with $p = 0.1$ as we decrease $m$. At **b** $m = 2.98$, Weyl nodes are created pairwise at $\mathbf{k}^+ = (0, 0, k_z^+)$ in $H_+$ sector. When $m$ becomes smaller, the Weyl nodes move in momentum space (see Fig. 3.4b). At **c** $m = 2.9$, the Weyl nodes exists but they are away from the $k_y = 0$ plane. Eventually, at **d** Weyl Semimetals and Spinless $Z_2$ Magnetic Topological Crystalline Insulators with Glide Symmetry $m = 2.82$, Weyl nodes are annihilated in pair at $\mathbf{k}^- = (0, 0, k_z^-)$. **e–j** Surface Fermi surface at $E = 0$ around the origin for the (100) surface. In **e–g** we fix $p = 0.1$, and change the parameter $m$ as **e** $m = 2.9$ (spinless WSM), **f** $m = 2.85$ (spinless WSM), and **g** $m = 2.8$ (TCI). Similarly, in **h–j** we fix $p = 0.4$, and change the parameter $m$ as **h** $m = 2.9$ (spinless WSM), **i** $m = 2.4$ (spinless WSM), and **j** $m = 2.2$ (TCI). A surface Dirac cone in the TCI phase appears in **g** and **j**, and the surface Fermi arcs in the spinless WSM phase appear in **e**, **f**, **h**, and **i**

both belonging to the $H_{\pm}$ sectors (i.e., $G_y = \pm e^{-ik_z/2}$). These closings of the bulk gap at $\mathbf{k}^{\pm}$ for $m = m^{\pm}$ in the $H_{\pm}$ sectors on the glide plane correspond to the pair creation or pair annihilation of Weyl nodes on the glide plane.

We compute the trajectory of the Weyl nodes by changing the parameter $m$ for several cases for the NI-WSM-TCI phase transitions. Here we set the parameters $t_0 = t_0' = t_0'' = t = t' = t'' = 1$. First, let us consider $\phi = 0.4$, $p = 0.1$ and decrease $m$ from the NI phase to the TCI phase. The overall trajectory of the Weyl nodes is illustrated in Fig. 3.4a, and the corresponding band structures along the $k_x = k_y = 0$ line are shown in Fig. 3.3b–d. The pair creation of two Weyl nodes occurs at $\mathbf{k}^{+}$, and they migrate in momentum space. The two Weyl nodes are eventually annihilated at $\mathbf{k}^{-}$ pairwise, and the system goes into the TCI. Hence these values of $m^{\pm}$ gives phase boundaries between the WSM and the TCI/NI phases shown as the curves ($A_{\pm}$) in the phase diagram Fig. 3.3a, given by $m^{\pm}, \mathbf{k}^{\pm}$.

Another type of trajectories appears when we set $\phi = 0.4$, $p = 0.4$. At these parameters, as $m$ decreases, the trajectory of the Weyl nodes is shown in Fig. 3.4b. Pair creations of Weyl nodes occur at two points $\tilde{\mathbf{k}}^{\pm} = (0, \tilde{k}_y^{\pm}, \tilde{k}_z)$, $m = m_0$ outside of the glide plane simultaneously, where $t_0' t''^2 \sin(\tilde{k}_z - \phi) = 4 t_0'' t'^2 \sin \tilde{k}_z \cos \tilde{k}_y^{\pm}$. The four Weyl nodes then migrate in momentum space. Thereafter, a pair annihilation of a pair of Weyl nodes firstly occurs at $\mathbf{k}^{+}, m = m^{+}$. The remaining pair of Weyl nodes is then annihilated at $\mathbf{k}^{-}, m = m^{-}$, and the system becomes the TCI phase. The pair creation of Weyl nodes at $\tilde{\mathbf{k}}^{\pm}$ outside of the glide plane leads to a phase boundary (B) in Fig. 3.3a. In these two cases illustrated in Fig. 3.4a, b, the two events of the pair creation/annihilation on the glide plane occur in the opposite glide sectors. It is hence consistent with our scenario for the changing of the glide-$Z_2$ invariant in Sect. 3.1.

### 3.2.3 Evolution of the Surface States

To exhibit evolution of the surface states in the above cases, we numerically compute band structure for a slab geometry with (100) surfaces that the glide symmetry $\hat{G}_y$ is preserved. Figure 3.3e–j show the Fermi surfaces of a slab at $E = 0$ for several values of the parameter $m$ with $p = 0.1, \phi = 0.4$ [(e)–(g)], and $p = 0.4, \phi = 0.4$ [(h)–(j)]. The system with $p = 0.1$ is in the spinless WSM for (e) $m = 2.9$ and (f) $m = 2.85$, with two Weyl nodes, and a single Fermi arc connects the projections of these Weyl nodes. By decreasing $m$, the two Weyl nodes are annihilated pairwise, and the system runs into the TCI phase. In (g) $m = 2.8$, the surface states become a Dirac cone in this TCI phase.

On the other hand, the system with $p = 0.4$ is in the spinless WSM with four Weyl nodes for (h) $m = 2.9$, and there are two Fermi arcs. By decreasing $m$, a pair of Weyl nodes is firstly annihilated at $\mathbf{k}^{+}$, and two Weyl nodes remain, resulting in a single Fermi arc as shown in (i) for $m = 2.4$. Later, the remaining pair is also annihilated at $\mathbf{k}^{-}$ and the system runs into the TCI. The surface Dirac cone for the

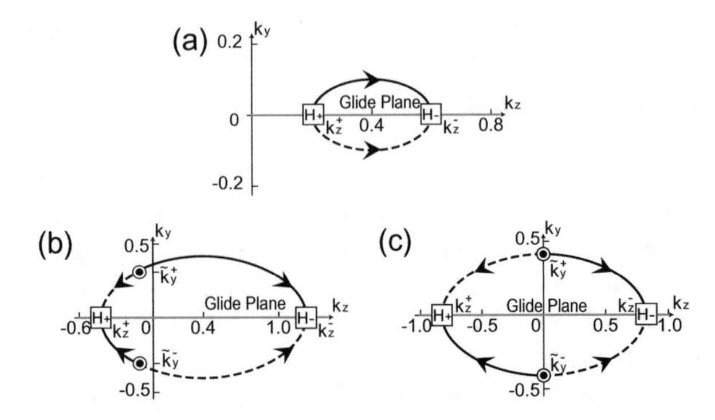


**Fig. 3.4** Numerical results for the trajectories of the Weyl nodes for our model of Eq. (3.16) for $t_0 = t_0' = t_0'' = t = t' = t'' = 1$ with decreasing $m$. $H_\pm$ on the glide plane ($k_y = 0$) represent glide sectors $G_y = \pm e^{-ik_z/2}$, respectively. **a** For $\phi = 0.4$, $p = 0.1$, a pair of Weyl nodes is created on the glide plane in the glide sector $H_+$ and eventually annihilated on the glide plane in the glide sector $H_-$. **b** For $\phi = 0.4$, $p = 0.4$, two pair creations occur simultaneously outside of the glide plane, and then the four Weyl nodes are annihilated in the glide sectors $H_+$ and $H_-$. **c** For $\phi = 0$, $p = 0.4$, because of the $C_{2x}$ symmetry, the trajectory is symmetrical with respect to the origin. In **a–c** the trajectories are symmetric with respect to the glide plane, with the sign change of the monopole charges

TCI phase at $m = 2.2$ is shown in Fig. 3.3j. Hence all the Weyl nodes are annihilated by going from the WSM phase to the TCI phase, and the two Fermi arcs evolve into the surface Dirac cone. Therefore we have seen that the surface Fermi arcs in the spinless WSM evolve into the surface Dirac cone in the TCI.

### 3.2.4 $\phi = 0$ *Case: Additional* $C_{2x}$ *Symmetry*

When $\phi = 0$, since the system has $C_{2x}$ symmetry in addition to the glide symmetry, the trajectory becomes symmetric with respect to the origin (see Fig. 3.4c). The two pair creations occur simultaneously, and so do the pair annihilations. In the present case, there are always four Weyl nodes within the WSM phase, and the phase boundaries ($A_+$) and ($A_-$) become identical. Figure 3.5a shows the phase diagram for $\phi = 0$ in the $m$-$p$ plane. Figure 3.5b–d are the surface Fermi surfaces at $E = 0$ on the (100) surfaces, where we fix $p = 0.4$ and switch the value of $m$. For (b) $m = 2.8$ and (c) $m = 2.7$, the system is in the spinless WSM, and the surface Fermi arcs are symmetric with respect to the origin. At (d) $m = 2.65$, the system is in the TCI phase, and the surface states become a Dirac cone. Therefore, as we go from the WSM phase to the TCI phase, the two Fermi arcs evolve into the surface Dirac cone. Note that the states in this case are always $C_{2x}$-symmetric and glide-symmetric.

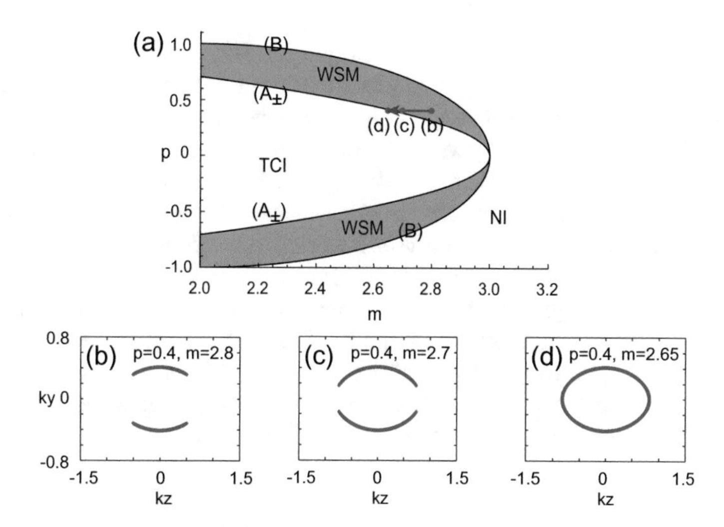


**Fig. 3.5** **a** Phase diagram in the $m$-$p$ plane for our model of Eq. (3.16) with $t_0 = t_0' = t_0'' = t = t' = t'' = 1$, $p = 0.4$, and $\phi = 0$. In this case, the system possesses $C_{2x}$ symmetry. **b–d** Surface Fermi surface at $E = 0$ for the (100) surface with the values of $m$ taken as **b** $m = 2.8$ (spinless WSM), **c** $m = 2.7$ (spinless WSM), and **d** $m = 2.65$ (TCI). A surface Dirac cone in the TCI phase appears in **d**, and the surface Fermi arcs in the spinless WSM phase appear in **b** and **c**

## 3.3 Conclusion and Discussion

The main conclusion in this chapter is that we showed that the spinless Weyl semimetal (WSM) phase emerges between topological crystalline insulator (TCI) and normal insulator (NI) phases in glide-symmetric systems in the absence of time-reversal symmetry. We first construct a theory for the TCI-NI phase transition with glide symmetry, and show a generic phase diagram involving the spinless WSM phase using the effective model. The trajectory of the Weyl nodes within the WSM phase governs the change of the $Z_2$ topological invariant. Especially, when the pair creation and annihilation of the Weyl nodes occur in the sectors with opposite signs of glide eigenvalues, the $Z_2$ topological invariant changes between the two sides of the WSM phase. These scenarios are supported by our spinless tight-binding model on a three-dimensional orthorhombic lattice with glide symmetry. In this model, we also show that surface Fermi arcs in the spinless WSM phase evolve into a surface Dirac cone in the TCI phase.

Let us compare the glide-symmetric spinless TCIs and the topological insulators (TIs). In order to accompany the WSM phase in the topological phase transition between a TI and a NI, it is necessary to break the spatial inversion symmetry [3, 5], provided the time-reversal symmetry is preserved. On the other hand, our results in the present chapter show that the WSM phase always appears between the spinless glide-symmetric TCI and the NI.

Materials search of the glide-symmetric spinless TCI would be a promising and interesting topic for first-principle calculations, because it does not suffer from the constraint of strong spin-orbit coupling, as has been the case for the TIs. For candidate spinless systems, it is necessary to break time-reversal symmetry but to preserve the glide symmetry. Such candidate materials might be found among spinless insulators coupled with localized spin systems or spinless insulators in a magnetic field, and search for such systems is left as a future work. As is similar to the spinless TCI for electrons, bosonic systems with glide symmetry can also support nontrivial $Z_2$ topological invariant, giving rise to a surface mode within a certain band gap. In this context, a theoretical proposal for the glide-$Z_2$ topological phase is reported in photonic crystal systems [6], on which we will discuss in detail in Chap. 6. From the results in the present chapter, it is also promising to discover new spinless WSM materials from such kind of glide-symmetric TCI materials.

## References


1. Fang C, Fu L (2015) Phys Rev B 91:161105(R). https://doi.org/10.1103/PhysRevB.91.161105
2. Shiozaki K, Sato M, Gomi K (2015) Phys Rev B 91:155120. https://doi.org/10.1103/PhysRevB.91.155120
3. Murakami S (2007) New J Phys 9(9):356
4. Murakami S, Kuga SI (2008) Phys Rev B 78:165313. https://doi.org/10.1103/PhysRevB.78.165313
5. Okugawa R, Murakami S (2014) Phys Rev B 89:235315. https://doi.org/10.1103/PhysRevB.89.235315
6. Lu L, Fang C, Fu L, Johnson SG, Joannopoulos JD, Soljačić M (2016) Nat Phys 12(4):337

# Chapter 4
# Interplay of Glide-Symmetric $Z_2$ Magnetic Topological Crystalline Insulators and Symmetry: Inversion Symmetry and Nonprimitive Lattice

It is known that three-dimensional (3D) magnetic systems with glide symmetry can be characterized by a $Z_2$ topological invariant together with the Chern number associated with the normal vector of the glide plane, and they are expressed in terms of integrals of the Berry curvature. In the present chapter, we study the fate of this topological invariant when inversion symmetry is added while time-reversal symmetry (TRS) is not enforced. There are two ways to add inversion symmetry to the space group No. 7 having glide symmetry only, leading to the space groups No. 13 and No. 14. In the space group No. 13, we find that the glide-$Z_2$ invariant is expressed solely from the irreducible representations (irreps) at high-symmetry points in $k$-space. It constitutes the $\mathbb{Z}_2 \times \mathbb{Z}_2$ symmetry-based indicators for this space group, together with another $\mathbb{Z}_2$ representing the Chern number modulo 2. In the space group No. 14, we find that the symmetry-based indicator $\mathbb{Z}_2$ is given by a combination of the glide-$Z_2$ invariant and the Chern number. Thus, in the space group No. 14, from the irreps at high-symmetry points we can only know possible combinations of the glide-$Z_2$ invariant and the Chern number, but in order to know each value of these topological numbers, we should calculate integrals of the Berry curvature. Moreover, we show that in both cases, the symmetry-based indicator $\mathbb{Z}_4$ for centrosymmetric systems leading to the higher-order topological insulators (TIs) is directly related to the glide-$Z_2$ invariant and the Chern number.

We also investigate topological invariants for glide-symmetric systems with non-primitive lattice with and without inversion symmetry, i.e., in the space groups No. 9 and No. 15. Meanwhile, in nonprimitive lattices, a half of the reciprocal vectors may not be invariant under the glide operations, and the formulas of topological invariants should be altered. In the space group No. 15, we find that the glide-$Z_2$ invariant is expressed solely from the irreps at high-symmetry points in $k$-space. It constitutes the $\mathbb{Z}_2 \times \mathbb{Z}_2$ symmetry-based indicators for this space group, together with another $\mathbb{Z}_2$ representing the Chern number modulo 2, similar to the space group No. 13. To this end, our approaches are twofold: one is a direct derivation of new formulas of the glide-$Z_2$ invariant on nonprimitive lattices via the space group No. 9, and the other is exploiting compatibility relations with the space group No. 13. Finally, we show that the symmetry-based indicator $\mathbb{Z}_4$ for centrosymmetric systems leading to the



H. Kim, *Glide-Symmetric $Z_2$ Magnetic Topological Crystalline Insulators*,
Springer Theses, https://doi.org/10.1007/978-981-16-9077-8_4

higher-order TIs is also directly related to the glide-$Z_2$ invariant in the space group No. 15.

Until the end of this thesis, we consider the type-I magnetic space groups, whose symbols are the same as those for the corresponding crystallographic space groups, because they do not contain time-reversal. For example, the space group No. 7 in this thesis refers to the type-I magnetic space group G $= Pc$ (magnetic space group 7.24 in the notation of Bilbao Crystallographic Server [1]) which consists of the same unitary operations as that of the space group No. 7. In the present and the following chapters, we mainly study the space groups No. 7, No. 9, No. 13, No. 14, and No. 15, which are the type-I magnetic space groups 7.24, 9.37, 13.65, 14.75, and 15.85, respectively. In addition, henceforth we abbreviate a space group to its number assigned in Ref. [2] in bold italic font. Namely, ***7***, ***9***, ***13***, ***14***, and ***15*** refer the type-I magnetic space groups 7.24, 9.37, 13.65, 14.75, and 15.85, which are the same to those crystallographic space groups No. 7, No. 9, No. 13, No. 14, and No. 15 [2], respectively.

## 4.1 Preliminaries

In this section, we will review basic properties of the glide-symmetric $Z_2$ magnetic topological crystalline insulator (TCI) [3, 4]. We start by introducing the $Z_2$ topological invariant characterizing the glide-symmetric system in the absence of TRS. We also introduce the previous works on topological phases for ***7***, ***9***, ***13***, ***14***, and ***15***. We finally give useful quantities at the end of this section used throughout the present chapter.

### *4.1.1 Previous Works on Topological Phases for Glide-Symmetric Systems with and Without Inversion Symmetry in Class A*

We begin with considering a 3D magnetic system with a glide operation

$$\hat{G}_y = \left\{ M_y \middle| \frac{1}{2}\hat{\mathbf{z}} \right\}, \tag{4.1}$$

where $M_y$ represents the mirror reflection with respect to the $xz$ plane, and $\hat{\mathbf{z}}$ is a unit vector along the $z$ axis. This space group is ***7*** ($Pc$). The Bloch Hamiltonian $\mathcal{H}(\mathbf{k})$ for the glide-symmetric system satisfies the equation

$$G_y(k_z)\mathcal{H}(k_x, k_y, k_z)G_y(k_z)^{-1} = \mathcal{H}(k_x, -k_y, k_z). \tag{4.2}$$

**Table 4.1** Summary of the generators, the $K$-groups, the symmetry-based indicators and the $E_\infty^{2,0}$ terms in the Atiyah-Hirzebruch spectral sequence for ***7***, ***9***, ***13***, ***14***, and ***15***. The $E_\infty^{2,0}$ terms represent the existence of topological invariants whose definitions involve an integral of Berry curvature on a two-dimensional subspace in the Brillouin zone. Here, $\mathbb{Z}^2 \equiv \mathbb{Z} \times \mathbb{Z}$ and $\mathbb{Z}_2 \equiv \mathbb{Z}/2\mathbb{Z}$

| | ***7*** $(Pc)$ | ***9***$(Cc)$ | ***13*** $(P2/c)$ | ***14*** $(P2_1/c)$ | ***15*** $(C2/c)$ |
|---|---|---|---|---|---|
| Generators | $\{M_y\|00\frac{1}{2}\}$ | $\{M_y\|00\frac{1}{2}\}$ | $\{I\|000\}$, $\{C_{2y}\|00\frac{1}{2}\}$ | $\{I\|000\}$, $\{C_{2y}\|0\frac{1}{2}\frac{1}{2}\}$ | $\{I\|000\}$, $\{C_{2y}\|00\frac{1}{2}\}$ |
| $K$-group | $\mathbb{Z}^2 \times \mathbb{Z}_2$ | $\mathbb{Z}^2 \times \mathbb{Z}_2$ | $\mathbb{Z}^8$ | $\mathbb{Z}^6$ | $\mathbb{Z}^7$ |
| Symmetry indicators | N/A | N/A | $\mathbb{Z}_2 \times \mathbb{Z}_2$ | $\mathbb{Z}_2$ | $\mathbb{Z}_2 \times \mathbb{Z}_2$ |
| $E_\infty^{2,0}$ | $\mathbb{Z} \times \mathbb{Z}_2$ | $\mathbb{Z} \times \mathbb{Z}_2$ | $\mathbb{Z}$ | $\mathbb{Z}$ | $\mathbb{Z}$ |

Then, we can define the $Z_2$ topological invariant for such glide-symmetric systems,

$$\begin{aligned}\nu = \frac{1}{2\pi}\Bigg[&-2\int_{-\pi}^{\pi} dk_x \mathrm{tr}\mathcal{A}_x^+(k_x, 0, -\pi) + 2\int_{-\pi}^{\pi} dk_x \mathrm{tr}\mathcal{A}_x^+(k_x, \pi, -\pi) \\ &+ \int_{-\pi}^{\pi} dk_x \int_{-\pi}^{\pi} dk_z \mathrm{tr}\mathcal{F}_{zx}^-(k_x, 0, k_z) - \int_{-\pi}^{\pi} dk_x \int_{-\pi}^{\pi} dk_z \mathrm{tr}\mathcal{F}_{zx}^-(k_x, \pi, k_z) \\ &+ \int_{0}^{\pi} dk_y \int_{-\pi}^{\pi} dk_x \mathrm{tr}\mathcal{F}_{xy}(k_x, k_y, -\pi)\Bigg] \pmod 2. \end{aligned} \tag{4.3}$$

The detail of the formalism is discussed in Sect. 3.1.1.

Here we consider what happens to the glide-$Z_2$ invariant, when inversion symmetry is added. The space group ***7***$(Pc)$, i.e., the space group with the glide symmetry only, becomes either ***13***$(P2/c)$ or ***14***$(P2_1/c)$ by adding inversion symmetry. The difference between ***13*** and ***14*** comes from the position of the inversion center; namely, while the inversion center in ***13*** is within the glide plane, that in ***14*** is not. The resulting space group ***13*** has twofold $(C_2)$ rotational symmetry and the space group ***14*** has $C_2$ screw symmetry.

In a nonprimitive lattice system, the fact that the length of its primitive lattice vectors is equal or less than the length of its conventional primitive lattice vectors yields a smaller unit cell than the conventional unit cell. This makes the Brillouin zone (BZ) of a nonprimitive lattice system bigger than the conventional BZ, and obviously high-symmetry points, lines, and planes should be changed. Therefore, the glide-$Z_2$ invariant $\nu$ in Eq. (4.3) should be altered in nonprimitive lattice systems. To figure this out, we consider glide-symmetric systems in base-centered lattice with and without inversion symmetry, i.e., in ***9***$(Cc)$ and ***15***$(C2/c)$.

For these five space groups, the $K$-groups, the classifications of insulators including atomic ones, are known to be $\mathbb{Z}^2 \times \mathbb{Z}_2$ (***7*** and ***9***), $\mathbb{Z}^8$ (***13***), $\mathbb{Z}^6$ (***14***), and $\mathbb{Z}^8$ (***15***) [5] (Table 4.1). The symmetry-based indicators, which are defined as the quotients of the $K$-groups by the abelian groups generated by atomic insulators, are trivial (***7*** and ***9***), $\mathbb{Z}_2 \times \mathbb{Z}_2$ (***13*** and ***15***), and $\mathbb{Z}_2$ (***14***) [6, 7] (Table 4.1). On the other hand, the topological invariants including the integral of Berry curvature on a 2D subspace of the BZ, which are classified by the $E_\infty^{2,0}$ terms in the Atiyah-Hirzebruch spectral sequence based on the $K$-theory, are classified by $\mathbb{Z} \times \mathbb{Z}_2$ (*7*), $\mathbb{Z} \times \mathbb{Z}_2$ (*9*), $\mathbb{Z}$ (*13*), $\mathbb{Z}$ (*14*), and $\mathbb{Z}$ (*15*) [5] (Table 4.1). However, physical meaning of $\mathbb{Z}^n$ and $\mathbb{Z}_n$ factors in Table 4.1 is not obvious. The recent work in Ref. [8] gives physical interpretations of symmetry-based indicators in class A by analyzing those explicit formulas in several key space groups. Nonetheless, from the symmetry-based indicators, one can neither study properties of topological invariants which cannot be expressed by symmetry-based indicators nor know how those topological invariants evolve when another symmetry is added and the space group becomes its supergroup.

### 4.1.2 Brief Review of Sewing Matrices and Monodromy

We exploit several convenient quantities argued in Ref. [9] for evaluating the integral of the Berry curvature in the presence of additional $C_2$ rotational symmetry or $C_2$ screw symmetry.

First, we define a sewing matrix $w_{mn}(\mathbf{k})$. The Bloch Hamiltonian $\mathcal{H}(\mathbf{k})$ having a point group symmetry $R$ satisfies $\hat{R}\mathcal{H}(\mathbf{k}) = \mathcal{H}(R\mathbf{k})\hat{R}$, where $\hat{R}$ is an operator corresponding to the point-group operation $R$, and $R\mathbf{k}$ denotes a wavevector transformed from $\mathbf{k}$ by $\hat{R}$. From the operator $\hat{R}$ which acts on the Bloch wavefunction $|\psi_n(\mathbf{k})\rangle = e^{i\mathbf{k}\cdot\mathbf{r}}|u_n(\mathbf{k})\rangle$, we define a corresponding operator $\tilde{R}$, which acts on the cell-periodic eigenstates $|u_n(\mathbf{k})\rangle$. Then, we define a unitary matrix

$$w_{mn}(\mathbf{k}) \equiv \langle u_m(R\mathbf{k})|\tilde{R}|u_n(\mathbf{k})\rangle, \tag{4.4}$$

where $m$ and $n$ run over the occupied states. At the wavevector $\mathbf{k}$ invariant under $R$, the Hamiltonian $\mathcal{H}(\mathbf{k})$ commutes with $\hat{R}$, so that we can find common eigenstates of the operators $\mathcal{H}(\mathbf{k})$ and $\tilde{R}$. Here, the sewing matrix $w_{mn}$ should be diagonal at such a high-symmetry point $\mathbf{k}_i$,

$$w_{mn}(\mathbf{k}_i) = R_m(\mathbf{k}_i)\delta_{mn}, \tag{4.5}$$

where $R_m$ is the eigenvalue of $\hat{R}$ for the $m$-th band. Therefore, the determinant of $w_{mn}$ is given by the product of the eigenvalues of occupied eigenstates at $\mathbf{k}_i$,

$$\det[w(\mathbf{k}_i)] = \prod_{n\in\mathrm{occ}} R_n(\mathbf{k}_i). \tag{4.6}$$

This quantity is gauge invariant because the determinant does not depend on the choice of basis.

Apart from the Berry phase, we also define a path-ordered Berry phase,

$$\mathcal{U}_{\mathbf{k}_1\mathbf{k}_2} = \mathcal{P}\exp\left[i\int_{\mathbf{k}_1}^{\mathbf{k}_2} \mathbf{A}(\mathbf{k})\cdot d\mathbf{k}\right], \tag{4.7}$$

evaluated in the subspace of the occupied states, along the path from $\mathbf{k}_1$ to $\mathbf{k}_2$, which is not necessarily being a closed path.

We also define an orbital-space operator

$$\tilde{U}_{\mathbf{k}_1\mathbf{k}_2} = \sum_{i,j\in\text{occ}} \left(\mathcal{U}_{\mathbf{k}_1\mathbf{k}_2}\right)_{ij} |u_i(\mathbf{k}_1)\rangle\langle u_j(\mathbf{k}_2)| \tag{4.8}$$

$$= \mathcal{P}\left[i\int_{\mathbf{k}_1}^{\mathbf{k}_2} \tilde{P}_{\mathbf{k}}\partial_{\mathbf{k}}\tilde{P}_{\mathbf{k}}\cdot d\mathbf{k}\right], \tag{4.9}$$

where $\tilde{P}_{\mathbf{k}} = \sum_{i\in\text{occ}}|u_i(\mathbf{k})\rangle\langle u_i(\mathbf{k})|$ is the projector onto the occupied subspace at $\mathbf{k}$. Unlike $\mathcal{U}_{\mathbf{k}_1\mathbf{k}_2}$, this operator is transformed as

$$\tilde{R}\tilde{U}_{\mathbf{k}_1\mathbf{k}_2}\tilde{R}^{-1} = \tilde{U}_{R\mathbf{k}_1 R\mathbf{k}_2}, \tag{4.10}$$

when the Hamiltonian is invariant under a point group $R$, since $\tilde{U}_{\mathbf{k}_1\mathbf{k}_2}$ is gauge invariant.

## 4.2 Redefinition of the Glide-$Z_2$ Invariant

In this section we revisit the definition of the glide-$Z_2$ invariant $\nu$. First we show how this topological invariant $\nu$ depends on the choice of the glide plane. We then introduce another topological invariant $\delta_g$ for glide-symmetric systems from the layer construction similarly to Ref. [10]. To achieve agreement between the two glide topological invariants $\nu$ and $\delta_g$, we show that it is better to change the definition of the glide-$Z_2$ invariant from $\nu$ to $\tilde{\nu} \equiv \nu + C_y$ (mod 2), where $C_y$ is the Chern number on the glide-invariant plane $k_y = 0$. The details will be discussed in Chap. 5. Henceforth, $C_y$ refers to the Chern number for each $k_y$-fixed plane. We will also explain that this redefinition does not affect physical properties of the glide-$Z_2$ invariant discussed in previous works [3, 4].

### 4.2.1 *Gauge Dependence of the Glide-$Z_2$ Invariant*

In systems with glide symmetry, we have two glide planes which are inequivalent under the lattice translation. For the glide operation given by Eq. (4.1), the glide planes are $y = 0$ and $y = \frac{1}{2}$. In such systems, when we take another glide operation given by

$$\hat{G}'_y \equiv \left\{ M'_y \middle| \frac{1}{2}\hat{\mathbf{z}} \right\}, \tag{4.11}$$

where $M'_y$ is a mirror reflection with respect to the $y = \frac{1}{2}$ plane whereas $M_y$ with respect to the $y = 0$ plane. Because $\hat{G}'_y = T_y \hat{G}_y$ where $T_y$ is a unit translation along the $y$ direction, this change of the glide operation switches the $g_+$ and $g_-$ sectors on the $k_y = \pi$ plane, while the glide sectors on the $k_y = 0$ plane remain intact. Therefore, the value of the glide-$Z_2$ invariant changes from $\nu$ to

$$\begin{aligned}\nu' = \frac{1}{2\pi}\bigg[&-2\int_{-\pi}^{\pi} dk_x \mathrm{tr}\mathcal{A}_x^+(k_x, 0, -\pi) + 2\int_{-\pi}^{\pi} dk_x \mathrm{tr}\mathcal{A}_x^-(k_x, \pi, -\pi)\\ &+\int_{-\pi}^{\pi} dk_x \int_{-\pi}^{\pi} dk_z \mathrm{tr}\mathcal{F}_{zx}^-(k_x, 0, k_z) - \int_{-\pi}^{\pi} dk_x \int_{-\pi}^{\pi} dk_z \mathrm{tr}\mathcal{F}_{zx}^+(k_x, \pi, k_z)\\ &+\int_{0}^{\pi} dk_y \int_{-\pi}^{\pi} dk_x \mathrm{tr}\mathcal{F}_{xy}(k_x, k_y, -\pi)\bigg] \pmod 2.\end{aligned} \tag{4.12}$$

Its difference from the original value $\nu$ is given by

$$\begin{aligned}\nu' - \nu = &-\frac{1}{2\pi}\int_{-\pi}^{\pi} dk_x \int_{-\pi}^{\pi} dk_z \mathrm{tr}\left(\mathcal{F}_{zx}^+(k_x, \pi, k_z) - \mathcal{F}_{zx}^-(k_x, \pi, k_z)\right)\\ &+\frac{1}{\pi}\left(\int_{-\pi}^{\pi} dk_x \mathrm{tr}\mathcal{A}_x^-(k_x, \pi, -\pi) - \int_{-\pi}^{\pi} dk_x \mathrm{tr}\mathcal{A}_x^+(k_x, \pi, -\pi)\right) \pmod 2.\end{aligned} \tag{4.13}$$

Because the last two terms can be rewritten as

$$\begin{aligned}&\int_{-\pi}^{\pi} dk_x \mathrm{tr}\mathcal{A}_x^-(k_x, \pi, -\pi) - \int_{-\pi}^{\pi} dk_x \mathrm{tr}\mathcal{A}_x^+(k_x, \pi, -\pi)\\ &= \int_{-\pi}^{\pi} dk_x \mathrm{tr}\mathcal{A}_x^+(k_x, \pi, \pi) - \int_{-\pi}^{\pi} dk_x \mathrm{tr}\mathcal{A}_x^+(k_x, \pi, -\pi)\\ &= \int_{-\pi}^{\pi} dk_x \int_{-\pi}^{\pi} dk_z \mathrm{tr}\mathcal{F}_{zx}^+(k_x, \pi, k_z),\end{aligned} \tag{4.14}$$

we get

$$\nu' - \nu = \frac{1}{2\pi}\int_{-\pi}^{\pi} dk_x \int_{-\pi}^{\pi} dk_z \mathrm{tr}\mathcal{F}_{zx}(k_x, \pi, k_z) = C_y \pmod 2, \tag{4.15}$$

where $C_y$ is the Chern number on the $k_z$–$k_x$ plane. Therefore, the value of the glide-$Z_2$ invariant changes with the change of the glide plane between $y = 0$ and $y = 1/2$, when $C_y$ is an odd integer.

### 4.2.2 Redefinition of the Glide-$Z_2$ Invariant

In systems with glide symmetry, one can introduce a topological invariant associated with the glide symmetry, from the viewpoint of the real-space layer construction. This is a different approach from the approach from the $k$-space topology, which led us to the glide-$Z_2$ invariant $\nu$. This approach of the layer construction has been formulated in Ref. [10] for systems with TRS. We here extend this theory of the layer construction to systems with glide symmetry without TRS. We will introduce a glide-invariant $\delta_g$ based on the geometry of the layers. The details of the definitions and calculations of $\delta_g$ will be discussed in Chap. 5.

By comparing this glide-invariant $\delta_g$ with the glide-$Z_2$ invariant $\nu$ in Eq. (4.3), we find that they are not equal. Nonetheless, if we redefine the glide-$Z_2$ invariant $\nu$ to be

$$\tilde{\nu} \equiv \nu + C_y \pmod 2, \tag{4.16}$$

it is equal to the glide invariant $\delta_g$:

$$\tilde{\nu} \equiv \delta_g. \tag{4.17}$$

One may wonder whether this redefinition may invalidate the bulk-boundary correspondence of the glide-$Z_2$ invariant $\nu$, meaning the number of helical surface states in the gap given in the previous works [3, 4]. Nevertheless, it is not the case, because in the arguments on the bulk-boundary correspondence in Refs. [3, 4], the Chern number $C_y$ is assumed to be equal to zero, in which case this redefinition does not alter the value of the glide-$Z_2$ invariant: $\nu = \tilde{\nu}$. Furthermore, the glide-$Z_2$ invariant does not depend on the choice of the glide plane, as follows from Eq. (4.15). On the other hand, we remark that when $C_y \neq 0$, the glide-$Z_2$ invariant may lose its meaning as the number of helical surface states, because its value is gauge dependent.

Thus the properties of the glide-$Z_2$ invariant in Refs. [3, 4] remain valid even with this redefinition, and henceforth we will adopt the redefined glide-$Z_2$ invariant (4.16). This redefinition is convenient when we compare the results with the layer construction, and is physically meaningful as we discuss later.

Here we give an expression for the new glide-$Z_2$ invariant $\tilde{\nu}$. From Eqs. (4.12) and (4.15), we conclude that $\tilde{\nu}$ is equal to $\nu'$. Thus, we get a formula for the new glide-$Z_2$ invariant

$$\begin{aligned}\tilde{\nu} = \frac{1}{2\pi}\Bigg[&-2\int_{-\pi}^{\pi} dk_x \mathrm{tr}\mathcal{A}_x^{+}(k_x, 0, -\pi) + 2\int_{-\pi}^{\pi} dk_x \mathrm{tr}\mathcal{A}_x^{-}(k_x, \pi, -\pi)\\ &+\int_{-\pi}^{\pi} dk_x \int_{-\pi}^{\pi} dk_z \mathrm{tr}\mathcal{F}_{zx}^{-}(k_x, 0, k_z) - \int_{-\pi}^{\pi} dk_x \int_{-\pi}^{\pi} dk_z \mathrm{tr}\mathcal{F}_{zx}^{+}(k_x, \pi, k_z)\\ &+\int_{0}^{\pi} dk_y \int_{-\pi}^{\pi} dk_x \mathrm{tr}\mathcal{F}_{xy}(k_x, k_y, -\pi)\Bigg] \pmod 2. \end{aligned}\tag{4.18}$$

Hereafter, we adopt this expression as the glide-$Z_2$ invariant.

## 4.3 Glide-Symmetric Magnetic Topological Crystalline Insulators with Inversion Symmetry in Primitive Lattice Systems

In the present section, we consider ***13*** and ***14***, which are realized by adding inversion symmetry to ***7***. Here we derive new formulas of the glide-$Z_2$ invariant (Eq. (4.18)) for ***13*** and ***14***. As explained in Sec. 4.1.1, ***13*** has $C_2$ rotational symmetry and the inversion center lies on the glide plane. Meanwhile, ***14*** has $C_2$ screw symmetry because the inversion center in ***14*** is not within the glide plane, leading to the property that the glide sector on the $k_y = \pi$ plane changes by $C_2$ screw rotation. This makes the calculation very different from that in ***13***. In deriving new formulas, we restrict ourselves to bulk-insulating systems. In particular, we exclude gapless phases with topological band-crossing points in bulk, such as Weyl semimetals [11, 12], because the glide-$Z_2$ invariant is ill-defined such a gapless phase.

### *4.3.1 Topological Invariants for Space Group 13*

We will show that in ***13***, the $Z_2$ topological invariant $\tilde{\nu}$ (mod 2) for the glide-symmetric system is given in terms of the irreps at high-symmetry points. For simplicity, here we consider the spinless case. Then, later we obtain

$$(-1)^{\tilde{\nu}} = \prod_{i \in \mathrm{occ}} \frac{\zeta_i^-(\Gamma)\zeta_i^+(C)}{\zeta_i^-(Y)\zeta_i^+(Z)}, \tag{4.19}$$

where $\zeta_i^{\pm}(=\pm 1)$ is an eigenvalue of the $C_2$ rotation for the eigenstates in the $g_{\pm}$ sector at high-symmetry points $\Gamma$, $Y$, $Z$, and $C$ shown in Fig. 4.1. In the following we show this formula by calculating $(-1)^{\tilde{\nu}}(= e^{i\pi\tilde{\nu}})$ from Eq. (4.18).

By adding an inversion $\hat{I} = \{I|\mathbf{0}\}$ around the origin to ***7***, we obtain ***13***. Then $C_2$ rotational symmetry around the axis $x = 0, z = 1/4$,

$$\hat{C}_2 = \hat{G}_y \hat{I} = \left\{ C_{2y} \middle| \frac{1}{2}\hat{\mathbf{z}} \right\}, \tag{4.20}$$

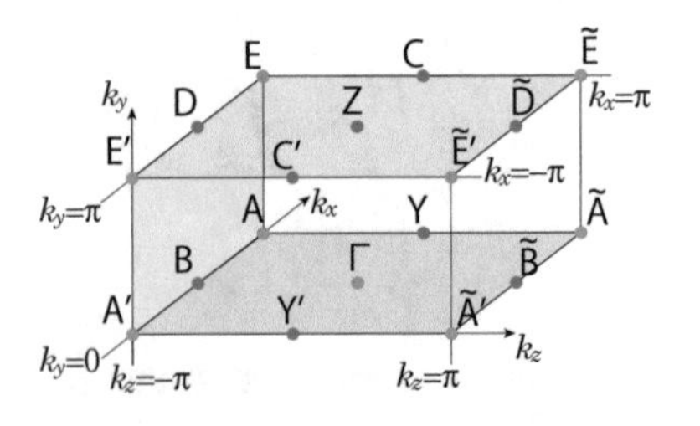


**Fig. 4.1** Upper half of the Brillouin zone of the monoclinic primitive lattice in ***7***, ***13***, and ***14***. $\Gamma$, $A$, $E$, $Y$, $Z$, $B$, $C$, and $D$ denote the high-symmetry points in ***13*** and ***14***

is also added to the symmetry operations. The key commutation relation between the glide operator and the $C_2$ rotational operator is given by

$$\hat{C}_2\hat{G}_y = \hat{G}_y\hat{C}_2\{E|\hat{\mathbf{z}}\}. \tag{4.21}$$

This will be used in the following discussion.

We now rewrite the formula of the glide-$Z_2$ invariant (4.18) with the help of additional symmetries. First, the Berry curvature on the $k_z = -\pi$ plane in Eq. (4.18) satisfies

$$\mathrm{tr}\mathcal{F}_{xy}(k_x, k_y, -\pi) = -\mathrm{tr}\mathcal{F}_{xy}(-k_x, k_y, \pi) = -\mathrm{tr}\mathcal{F}_{xy}(-k_x, k_y, -\pi), \tag{4.22}$$

owing to the $C_2$ symmetry. This immediately leads to

$$\int_0^\pi dk_y \oint_{-\pi}^\pi dk_x \mathrm{tr}\mathcal{F}_{xy}(k_x, k_y, -\pi) = 0. \tag{4.23}$$

Next, we address the Berry curvature on the glide-invariant planes $k_y = 0$ and $k_y = \pi$. Equation (4.21) indicates that the glide sector for wavefunctions within the glide-invariant planes is unchanged under the $C_2$ rotation. Therefore, the Berry curvatures $\mathcal{F}_{zx}^{\pm}(\mathbf{k})$ on those planes have to be an even function of $k_x$ and $k_z$,

$$\mathrm{tr}\mathcal{F}_{zx}^{\pm}(k_x, k_y, k_z) = \mathrm{tr}\mathcal{F}_{zx}^{\pm}(-k_x, k_y, -k_z), \tag{4.24}$$

where $k_y$ is either 0 or $\pi$.

Let us first consider the integral on the glide-invariant plane $k_y = 0$. From Eq. (4.24), we get

$$\int_{-\pi}^\pi dk_z \oint_{-\pi}^\pi dk_x \mathrm{tr}\mathcal{F}_{zx}^{-}(k_x, 0, k_z) = 2\int_0^\pi dk_z \oint_{-\pi}^\pi dk_x \mathrm{tr}\mathcal{F}_{zx}^{-}(k_x, 0, k_z). \tag{4.25}$$

Note that the $\hat{C}_2$ does not change the glide sector of the eigenstates within the $k_y = 0$ plane and we have taken the gauge to be periodic along the $k_x$ direction. Therefore, by using the Stokes' theorem on the $k_y = 0$ plane, we have

$$\begin{aligned}
&\exp\left[\frac{i}{2}\int_{-\pi}^\pi dk_z \oint_{-\pi}^\pi dk_x \mathrm{tr}\mathcal{F}_{zx}^{-}(k_x, 0, k_z) - i\oint_{-\pi}^\pi dk_x \mathrm{tr}\mathcal{A}_x^{+}(k_x, 0, -\pi)\right] \\
&= \exp\left[i\int_0^\pi dk_z \oint_{-\pi}^\pi dk_x \mathrm{tr}\mathcal{F}_{zx}^{-}(k_x, 0, k_z) - i\oint_{-\pi}^\pi dk_x \mathrm{tr}\mathcal{A}_x^{+}(k_x, 0, -\pi)\right] \\
&= \exp\left[-i\oint_{-\pi}^\pi dk_x \mathrm{tr}\mathcal{A}_x^{-}(k_x, 0, 0)\right].
\end{aligned} \tag{4.26}$$

We have used the relation of the branch cut, $\exp\left[i\oint_{-\pi}^\pi dk_x \mathrm{tr}\mathcal{A}_x^{+}(k_x, 0, -\pi)\right] = \exp\left[i\oint_{-\pi}^\pi dk_x \mathrm{tr}\mathcal{A}_x^{-}(k_x, 0, \pi)\right] \pmod{2\pi}$.

Similarly, the terms on the other glide-invariant plane $k_y = \pi$ are given by

$$\exp\left[-\frac{i}{2}\int_{-\pi}^{\pi} dk_z \oint_{-\pi}^{\pi} dk_x \mathrm{tr}\mathcal{F}_{zx}^{+}(k_x, \pi, k_z) + i\oint_{-\pi}^{\pi} dk_x \mathrm{tr}\mathcal{A}_x^{-}(k_x, \pi, -\pi)\right]$$
$$= \exp\left[i\oint_{-\pi}^{\pi} dk_x \mathrm{tr}\mathcal{A}_x^{+}(k_x, \pi, 0)\right]. \tag{4.27}$$

We note that Eqs. (4.26) and (4.27) are gauge invariant.

Consequently, from Eqs. (4.23), (4.26), and (4.27), we obtain a formula for the glide-$Z_2$ invariant (4.18) as

$$(-1)^{\tilde{\nu}} = \exp\left[-i\oint_{-\pi}^{\pi} dk_x \mathrm{tr}\mathcal{A}_x^{-}(k_x, 0, 0)\right]\exp\left[i\oint_{-\pi}^{\pi} dk_x \mathrm{tr}\mathcal{A}_x^{+}(k_x, \pi, 0)\right]$$
$$= \prod_{i\in\mathrm{occ}} \frac{\zeta_i^{-}(\Gamma)\zeta_i^{+}(C)}{\zeta_i^{-}(Y)\zeta_i^{+}(Z)}, \tag{4.28}$$

where $\zeta_i^{+}$ and $\zeta_i^{-}$ are the $C_2$ eigenvalues ($= \pm 1$) of the $i$-th occupied state in the $g_{+}$ and $g_{-}$ sectors, respectively, and $\Gamma$, $Y$, $Z$, and $C$ are the high-symmetry points on the plane $k_z = 0$ shown in Fig. 4.1. To show this formula we used the properties of the sewing matrix as introduced in Ref. [9] and its details are explained in Appendix 1.1.

At the four high-symmetry points $\Gamma$, $Y$, $Z$, and $C$, all the irreps are 1D, and those irreps are shown in Table 4.2. Therefore, an alternative expression for $\tilde{\nu}$ is

$$\tilde{\nu} = N_{\Gamma_2^{+}}(\Gamma) + N_{Y_2^{+}}(Y) + N_{C_2^{-}}(C) + N_{Z_2^{-}}(Z) \pmod 2, \tag{4.29}$$

where $N_R(P)$ is the number of occupied states at the high-symmetry point $P$ with an irrep $R$ for the $k$-group of ***13***.

Equations (4.28) and (4.29) explicitly depend on glide eigenvalues at high-symmetry points, and they look convention-dependent due to $4\pi$ periodicity of the glide eigenvalues. Nevertheless, they can be rewritten in terms of the parity eigenvalues and thus are independent of conventions. By using compatibility relations for the irreps of ***13***, we can rewrite the formula for the glide-$Z_2$ invariant $\tilde{\nu}$ as follows. We introduce the $z_4$ indicator for centrosymmetric systems

$$z_4 = \sum_{\mathbf{K}\in\mathrm{TRIM}} \frac{n_{\mathbf{K}}^{+} - n_{\mathbf{K}}^{-}}{2} \pmod 4, \tag{4.30}$$

where $n_{\mathbf{K}}^{+}$ and $n_{\mathbf{K}}^{-}$ are the number of occupied even-parity and odd-parity states at a time-reversal invariant momentum (TRIM) $\mathbf{K}$, and the sum is taken over the eight TRIMs.

As mentioned earlier, we are assuming insulating systems, and in particular we exclude the Weyl semimetal phase, which means that this $z_4$ indicator is always an even integer [8, 13]. Here we briefly explain why the $z_4$ indicator should be even.

**Table 4.2** Summary of irreducible representations of the little group and the numbers of irreducible representations for ***13*** satisfying the compatibility relations, where $a, b, m, x, y, z$ and $l$ are integers

| Seitz | $\{1\|t_1, t_2, t_3\}$ | $\{2_{010}\|0, 0, 1/2\}$ | $\{\bar{1}\|0, 0, 0\}$ | $\{m_{010}\|0, 0, 1/2\}$ | Number of irreps |
|---|---|---|---|---|---|
| Matrix presentation | $\begin{pmatrix} 1 & 0 & 0 & t_1 \\ 0 & 1 & 0 & t_2 \\ 0 & 0 & 1 & t_3 \end{pmatrix}$ | $\begin{pmatrix} -1 & 0 & 0 & 0 \\ 0 & 1 & 0 & 0 \\ 0 & 0 & -1 & 1/2 \end{pmatrix}$ | $\begin{pmatrix} -1 & 0 & 0 & 0 \\ 0 & -1 & 0 & 0 \\ 0 & 0 & -1 & 0 \end{pmatrix}$ | $\begin{pmatrix} 1 & 0 & 0 & 0 \\ 0 & -1 & 0 & 0 \\ 0 & 0 & 1 & 1/2 \end{pmatrix}$ | |
| $\Gamma_1^+$ | 1 | 1 | 1 | 1 | $a$ |
| $\Gamma_1^-$ | 1 | 1 | −1 | −1 | $b+m$ |
| $\Gamma_2^+$ | 1 | −1 | 1 | −1 | $a-m$ |
| $\Gamma_2^-$ | 1 | −1 | −1 | 1 | $b$ |
| $Y_1^+$ | $e^{i\pi t_1}$ | 1 | 1 | 1 | $a+x$ |
| $Y_1^-$ | $e^{i\pi t_1}$ | 1 | −1 | −1 | $b+y$ |
| $Y_2^+$ | $e^{i\pi t_1}$ | −1 | 1 | −1 | $a-y$ |
| $Y_2^-$ | $e^{i\pi t_1}$ | −1 | −1 | 1 | $b-x$ |
| $Z_1^+$ | $e^{i\pi t_2}$ | 1 | 1 | 1 | $a+z$ |
| $Z_1^-$ | $e^{i\pi t_2}$ | 1 | −1 | −1 | $b+m-z$ |
| $Z_2^+$ | $e^{i\pi t_2}$ | −1 | 1 | −1 | $a-m+z$ |
| $Z_2^-$ | $e^{i\pi t_2}$ | −1 | −1 | 1 | $b-z$ |
| $C_1^+$ | $e^{i\pi(t_1+t_2)}$ | 1 | 1 | 1 | $a+x+l$ |
| $C_1^-$ | $e^{i\pi(t_1+t_2)}$ | 1 | −1 | −1 | $b+y-l$ |
| $C_2^+$ | $e^{i\pi(t_1+t_2)}$ | −1 | 1 | −1 | $a-y+l$ |
| $C_2^-$ | $e^{i\pi(t_1+t_2)}$ | −1 | −1 | 1 | $b-x-l$ |
| $B_1$ | $e^{i\pi t_3}\sigma_0$ | $\sigma_x$ | $\sigma_z$ | $-i\sigma_y$ | $a+b$ |
| $A_1$ | $e^{i\pi(t_1+t_3)}\sigma_0$ | $\sigma_x$ | $\sigma_z$ | $-i\sigma_y$ | $a+b$ |
| $D_1$ | $e^{i\pi(t_2+t_3)}\sigma_0$ | $\sigma_x$ | $\sigma_z$ | $-i\sigma_y$ | $a+b$ |
| $E_1$ | $e^{i\pi(t_1+t_2+t_3)}\sigma_0$ | $\sigma_x$ | $\sigma_z$ | $-i\sigma_y$ | $a+b$ |

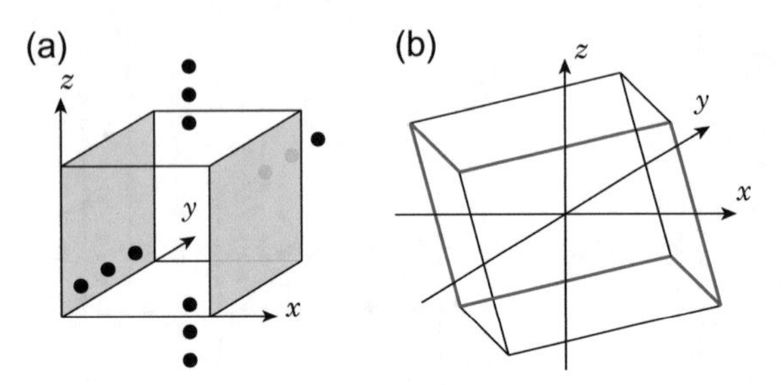


**Fig. 4.2** Illustration of **a** the surface states (blue planes) of the glide-$Z_2$ topological crystalline insulators and **b** the hinge states (red lines) of the higher-order topological insulator ensured by inversion symmetry

When the $z_4$ indicator is odd, which means that the parity of the Chern number on the $k_y = 0$ plane and that on the $k_y = \pi$ plane are not equal, the gap should close somewhere between these planes, and the system is no longer an insulator [13]. In such case, the gap closes at Weyl nodes, which are apices of the Dirac cones and carry topological charges characterized by the monopole density $\rho(\mathbf{k}) \equiv \frac{1}{2\pi}\nabla_{\mathbf{k}} \cdot \mathbf{\Omega}(\mathbf{k})$ [11, 12], where $\mathbf{\Omega}(\mathbf{k})$ is the summation of the Berry curvature for all of the occupied states. Therefore, the system is no longer an insulator, which is to be excluded here. Thus, the $z_4$ indicator is always an even integer and when it takes a value $z_4 \equiv 2 \pmod 4$ the system is a higher-order TI with gapless hinge states. Then from the compatibility relations for ***13***, we obtain

$$\tilde{\nu} \equiv \frac{z_4}{2} \pmod 2. \tag{4.31}$$

To show this relation, we use the compatibility relation determining the relations among the numbers of irreps $R$ at high-symmetry points summarized in the rightmost column of Table 4.2. Namely, within the abelian group generated by the irreps at high-symmetry points, the integer parameters $a, b, \ldots$ in these numbers corresponds to the generators of this abelian group satisfying the compatibility relations. Therefore, the number of the parameters is equal to the dimension $d_{\mathrm{BS}}$ in $\mathbb{Z}^{d_{\mathrm{BS}}}$ in Ref. [6], and we have $d_{\mathrm{BS}} = 7$ for ***13*** which corresponds to the $k$-group in Table 4.1. In ***13***, from the Table 4.2, the glide-$Z_2$ invariant $\tilde{\nu}$ is equal to $m + x + y + z + l \pmod 2$. On the other hand, the $\mathbb{Z}_4$ index $z_4$ is calculated by counting the number of odd-parity states at all the TRIMs, and we get $z_4 \equiv 2m - 2x + 2y - 2z - 2l = 2(m + x + y + z + l) = 2\tilde{\nu} \pmod 4$. Therefore, we verify that the glide-$Z_2$ invariant $\tilde{\nu} \pmod 2$ is equivalent to one half of the $\mathbb{Z}_4$ index $z_4$ for centrosymmetric systems, from the compatibility relation for irreps.

One may wonder how the surface states of the glide-$Z_2$ TCI and the hinge states of the higher-order TI ensured by inversion symmetry appear. In our case of Eq. (4.1), the (100) surface preserves the glide symmetry and the surface states of the glide-$Z_2$ TCI emerge on this plane (Fig. 4.2a). On the other hand, if one consider a rod or a cuboid configuration without (100) surfaces, the surface states disappear while the hinge states remain (Fig. 4.2b). We note that the hinge states are localized with a short penetration depth when the gap size of the surface states is large.

Since TRS is not assumed here, the Chern number $C_y$ on the $k_x$-$k_z$ plane, which is parallel to the glide plane, can be nonzero. On the other hand, the Chern numbers on the $k_x$-$k_y$ plane ($C_x$) and on the $k_y$-$k_z$ plane ($C_z$) identically vanish due to the $C_2$ rotational symmetry. As argued in Refs. [9, 13], the Chern number on the $k_x$-$k_z$ plane, which is perpendicular to the $\hat{C}_2$ axis, is calculated in terms of $C_2$ eigenvalues up to modulo 2. On the $k_y = 0$ plane, it is expressed as

$$(-1)^{C_y} = \prod_{i \in \text{occ}} \zeta_i(\Gamma)\zeta_i(Y)\zeta_i(B)\zeta_i(A), \tag{4.32}$$

where $\zeta_i$ is the $C_2$ eigenvalue of the $i$-th occupied state at the high-symmetry points, $\Gamma, Y, B$, and $A$ on the $k_y = 0$ plane shown in Fig. 4.1. In the same manner, the Chern number on the $k_y = \pi$ plane is calculated from the product of $C_2$ eigenvalues at $Z, C, D$, and $E$, and, it should be equal to $C_y$ calculated on the $k_y = 0$ plane, since we have assumed a presence of the bulk gap. Otherwise, if the Chern numbers calculated at $k_y = 0$ and at $k_y = \pi$ are not equal, the band gap vanishes somewhere between these planes [13], which occurs when $z_4$ is odd. On the other hand, this cannot occur if $z_4$ is even, i.e., $z_4 = 0 \pmod 2$.

### 4.3.2 Topological Invariants for 14

Next, we derive the glide-$Z_2$ invariant in ***14*** by adding inversion symmetry to ***7***. Here we consider the spinless case for simplicity. In this case, we consider glide and inversion symmetries,

$$\hat{G}_y = \left\{ M_y \middle| \frac{1}{2}\hat{\mathbf{y}} + \frac{1}{2}\hat{\mathbf{z}} \right\}, \quad \hat{I} = \{I|\mathbf{0}\}, \tag{4.33}$$

with its inversion center at the origin. It means that the glide plane is $y = 1/4$, and it does not contain the inversion center, unlike that in ***13***. Then we get $C_2$ screw symmetry

$$\hat{S}_y = \hat{G}_y \hat{I} = \left\{ C_{2y} \middle| \frac{1}{2}\hat{\mathbf{y}} + \frac{1}{2}\hat{\mathbf{z}} \right\}. \tag{4.34}$$

The commutation relation between the glide and the $C_2$ screw operators is given by

$$\hat{S}_y \hat{G}_y = \hat{G}_y \hat{S}_y \{E| - \hat{\mathbf{y}} + \hat{\mathbf{z}}\}. \tag{4.35}$$

Below, we mainly follow the same spirit as the previous subsection to derive a formula of the glide-$Z_2$ invariant for ***14***. The symmetry-based indicator is $\mathbb{Z}_2$ for ***14*** [6, 7]. In the following, we show that the indicator $\mathbb{Z}_2$ gives information on the combination of the glide-$Z_2$ invariant and the Chern number on the $k_z$-$k_x$ plane. Its formula here we find reads

$$C_y \in 2\mathbb{Z}, \quad (-1)^{\tilde{\nu}}(-1)^{C_y/2} = \prod_{i\in\mathrm{occ}} \frac{\eta_i^-(\Gamma)\eta_i^+(D)}{\eta_i^-(Y)\eta_i^+(E)}, \tag{4.36}$$

where $\eta_i^{\pm}$ is an eigenvalue of the $C_2$ screw operation in the $g_{\pm}$ sector at the high-symmetry points $\Gamma$, $Y$, $D$, and $E$. In the following we show this formula by calculating $(-1)^{\tilde{\nu}}(= e^{i\pi\tilde{\nu}})$ with $\tilde{\nu}$ given by Eq. (4.18). In the derivation, we should pay attention to the glide sectors and the branch cut.

Let us start with the $k_z = -\pi$ plane term in Eq. (4.18). The Berry curvature on the $xy$ plane with $k_z = -\pi$ should be an odd function of $k_x$ because of the $C_2$ screw symmetry. Therefore, we get

$$\int_0^{\pi} dk_y \oint_{-\pi}^{\pi} dk_x \mathrm{tr}\mathcal{F}_{xy}(k_x, k_y, -\pi) = 0, \tag{4.37}$$

as similar to Eq. (4.23) in ***13***.

On the $k_y = 0$ plane, because the $C_2$ screw merely behaves as $C_2$ rotation and it does not alter the glide sector, we have

$$\exp\left[\frac{i}{2}\int_{-\pi}^{\pi} dk_z \oint_{-\pi}^{\pi} dk_x \mathrm{tr}\mathcal{F}_{zx}^-(k_x, 0, k_z) - i\oint_{-\pi}^{\pi} dk_x \mathrm{tr}\mathcal{A}_x^+(k_x, 0, -\pi)\right]$$
$$= \exp\left[-i\oint_{-\pi}^{\pi} dk_x \mathrm{tr}\mathcal{A}_x^-(k_x, 0, 0)\right], \tag{4.38}$$

which is the same as Eq. (4.26) in ***13***.

On the other hand, we confront a difference from the case of ***13***, when we evaluate the integral over the other glide-invariant plane $k_y = \pi$. One cannot use the same trick as in ***13*** because the $C_2$ screw changes the glide sectors, leading to the relation $\mathcal{F}_{zx}^+(k_x, \pi, k_z) = \mathcal{F}_{zx}^-(-k_x, \pi, -k_z)$. Due to this relation, we find that the integral of the Berry curvature for the $g_+$ sector is a half of the integral of the total Berry curvature on the $k_y = \pi$ plane:

$$\exp\left[-\frac{i}{2}\int_{-\pi}^{\pi} dk_z \oint_{-\pi}^{\pi} dk_x \mathrm{tr}\mathcal{F}_{zx}^+(k_x, \pi, k_z) + i\oint_{-\pi}^{\pi} dk_x \mathrm{tr}\mathcal{A}_x^-(k_x, \pi, -\pi)\right]$$
$$= \exp\left[-\frac{i}{4}\oint_{-\pi}^{\pi} dk_z \oint_{-\pi}^{\pi} dk_x \mathrm{tr}\mathcal{F}_{zx}(k_x, \pi, k_z) + i\oint_{-\pi}^{\pi} dk_x \mathrm{tr}\mathcal{A}_x^-(k_x, \pi, -\pi)\right]$$
$$= (-i)^{C_y}\exp\left[i\oint_{-\pi}^{\pi} dk_x \mathrm{tr}\mathcal{A}_x^-(k_x, \pi, -\pi)\right]. \tag{4.39}$$

In the similar manner, noticing that

$$\begin{aligned}
&\exp\left[i\oint_{-\pi}^{\pi} dk_x \mathrm{tr}\mathcal{A}_x^-(k_x,\pi,-\pi)\right] \\
&=\exp\left[i\oint_{-\pi}^{\pi} dk_x \mathrm{tr}\mathcal{A}_x^+(k_x,\pi,\pi)\right] \\
&=\exp\left[i\int_{-\pi}^{\pi} dk_z \oint_{-\pi}^{\pi} dk_x \mathrm{tr}\mathcal{F}_{zx}^+ + i\oint_{-\pi}^{\pi} dk_x \mathrm{tr}\mathcal{A}_x^+(k_x,\pi,-\pi)\right] \\
&=(-1)^{C_y}\exp\left[i\oint_{-\pi}^{\pi} dk_x \mathrm{tr}\mathcal{A}_x^+(k_x,\pi,-\pi)\right],
\end{aligned} \tag{4.40}$$

we can rewrite the terms on the $k_y=\pi$ plane as

$$\begin{aligned}
&\exp\left[-\frac{i}{2}\int_{-\pi}^{\pi} dk_z \oint_{-\pi}^{\pi} dk_x \mathrm{tr}\mathcal{F}_{zx}^+(k_x,\pi,k_z) + i\oint_{-\pi}^{\pi} dk_x \mathrm{tr}\mathcal{A}_x^-(k_x,\pi,-\pi)\right] \\
&= i^{C_y}\exp\left[i\oint_{-\pi}^{\pi} dk_x \mathrm{tr}\mathcal{A}_x^+(k_x,\pi,-\pi)\right].
\end{aligned} \tag{4.41}$$

As a consequence, by combining Eqs. (4.37), (4.38), and (4.41), we recast the formula of the glide-$Z_2$ invariant into

$$\begin{aligned}
(-1)^{\tilde{\nu}} &= i^{C_y}\exp\left[-i\oint_{-\pi}^{\pi} dk_x \mathrm{tr}\mathcal{A}_x^-(k_x,0,0)\right]\exp\left[i\oint_{-\pi}^{\pi} dk_x \mathrm{tr}\mathcal{A}_x^+(k_x,\pi,-\pi)\right] \\
&= i^{C_y}\prod_{i\in\mathrm{occ}}\frac{\eta_i^-(\Gamma)\eta_i^+(D)}{\eta_i^-(Y)\eta_i^+(E)},
\end{aligned} \tag{4.42}$$

where $\prod_{i\in\mathrm{occ}}\eta_i^{\pm}(P)$ is the product of the $C_2$ screw eigenvalues over the occupied states for the $g_{\pm}$ sector at the high-symmetry point $P$. An alternative expression is

$$(-1)^{\tilde{\nu}}(-i)^{C_y} = \prod_{i\in\mathrm{occ}}\frac{\eta_i^-(\Gamma)\eta_i^+(D)}{\eta_i^-(Y)\eta_i^+(E)}. \tag{4.43}$$

Since the product of $\eta_i^{\pm}$ in Eq. (4.43) is equal to $\pm1$, the Chern number is found to be an even integer. Therefore, we have eventually shown that

$$C_y\in 2\mathbb{Z},\quad (-1)^{\tilde{\nu}}(-1)^{C_y/2} = \prod_{i\in\mathrm{occ}}\frac{\eta_i^-(\Gamma)\eta_i^+(D)}{\eta_i^-(Y)\eta_i^+(E)}. \tag{4.44}$$

This corresponds to the symmetry-based indicator $\mathbb{Z}_2$ for ***14***. Therefore, an alternative expression is

$$\tilde{\nu}+\frac{C_y}{2} = N_{\Gamma_2^+}(\Gamma)+N_{Y_2^+}(Y)+N_{D_2^-}(D)+N_{E_2^-}(E) \pmod 2. \tag{4.45}$$

The irreps are given in Table 4.3. Furthermore, from the compatibility relations, one can also show that the $z_4$ indicator for centrosymmetric systems is directly related to this symmetry-based indicator Eq. (4.45):

$$\tilde{\nu} + \frac{C_y}{2} = \frac{z_4}{2} \pmod 2. \tag{4.46}$$

Let us check the compatibility relations. The numbers of irreps $R$ at high-symmetry points are summarized in the rightmost column of Table 4.3. The integer parameters $a, b, \ldots$ in these numbers correspond to the generators of the abelian group generated by the irreps at high-symmetry points, and therefore the dimension of this abelian group is given by $d_{\mathrm{BS}} = 5$ for ***14*** [6] which corresponds to the $k$-group in Table 4.1. By using this Table and Eq. (4.45), we obtain $\tilde{\nu} + \frac{1}{2}C_y \equiv x + y + z \pmod 2$. On the other hand, the $\mathbb{Z}_4$ index for centrosymmetric systems is equal to $z_4 = -2(x + y + z) = 2(x + y + z) = 2(\tilde{\nu} + C_y/2) \pmod 4$ and Eq. (4.46) is shown. Thus the value of the symmetry-based indicator $\mathbb{Z}_2$ for ***14***, being equal to $z_4/2$, constrains possible combinations of the value of the glide-$Z_2$ invariant $\tilde{\nu}$ and that of the Chern number $C_y$.

## 4.4 Glide-Symmetric Magnetic Topological Crystalline Insulators in Nonprimitive Lattice Systems

In this section, we focus on glide-symmetric magnetic systems with a nonprimitive lattice. As we mentioned, the previous formula of the glide-$Z_2$ invariant does not apply to nonprimitive lattice systems since a half of the reciprocal vectors may not be invariant under the glide operation, and it should be altered. We first rewrite the glide-$Z_2$ invariant for ***9*** constituting a nonprimitive lattice with glide symmetry only. Then, we explore ***15*** having inversion symmetry.

### *4.4.1 Motivation*

Let us compare ***7*** ($Pc$) with ***9*** ($Cc$). While ***7*** constitutes a primitive lattice, ***9*** constitutes a $c$ base-centered lattice with glide symmetry in Eq. (4.1) only. The primitive and reciprocal lattice vectors for ***7*** and ***9*** are summarized in Table 4.4. Here we set the lattice constant to be unity for simplicity.

Obviously, the $k_y = 0$ plane is invariant under the glide symmetry from Eq. (4.2). From Eq. (4.2), the Bloch Hamiltonian is also invariant under the glide operator on the plane $k_y = \pi$ in BZ if the $k_y = \pi$ plane is perpendicular to one of the reciprocal primitive vectors. In such case, one can also calculate the Berry curvature for the glide subspaces associated with the glide-invariant plane $k_y = \pi$. Nevertheless, in nonprimitive lattices where the glide-invariant plane $k_y = \pi$ is not perpendicular to

**Table 4.3** Summary of irreducible representations of the little group and the numbers of irreducible representations for ***14*** satisfying the compatibility relations, where $a, b, x, y$ and $z$ are integers

| Seitz | $\{1\|t_1, t_2, t_3\}$ | $\{2_{010}\|0, 1/2, 1/2\}$ | $\{\bar{1}\|0, 0, 0\}$ | $\{m_{010}\|0, 1/2, 1/2\}$ | Number of irreps |
|---|---|---|---|---|---|
| Matrix presentation | $\begin{pmatrix} 1 & 0 & 0 & t_1 \\ 0 & 1 & 0 & t_2 \\ 0 & 0 & 1 & t_3 \end{pmatrix}$ | $\begin{pmatrix} -1 & 0 & 0 & 0 \\ 0 & 1 & 0 & 1/2 \\ 0 & 0 & -1 & 1/2 \end{pmatrix}$ | $\begin{pmatrix} -1 & 0 & 0 & 0 \\ 0 & -1 & 0 & 0 \\ 0 & 0 & -1 & 0 \end{pmatrix}$ | $\begin{pmatrix} 1 & 0 & 0 & 0 \\ 0 & -1 & 0 & 1/2 \\ 0 & 0 & 1 & 1/2 \end{pmatrix}$ | |
| $\Gamma_1^+$ | 1 | 1 | 1 | 1 | $a$ |
| $\Gamma_1^-$ | 1 | 1 | $-1$ | $-1$ | $b$ |
| $\Gamma_2^+$ | 1 | $-1$ | 1 | $-1$ | $a$ |
| $\Gamma_2^-$ | 1 | $-1$ | $-1$ | 1 | $b$ |
| $Y_1^+$ | $e^{i\pi t_1}$ | 1 | 1 | 1 | $a+x$ |
| $Y_1^-$ | $e^{i\pi t_1}$ | 1 | $-1$ | $-1$ | $b-x$ |
| $Y_2^+$ | $e^{i\pi t_1}$ | $-1$ | 1 | $-1$ | $a+x$ |
| $Y_2^-$ | $e^{i\pi t_1}$ | $-1$ | $-1$ | 1 | $b-x$ |
| $Z_1$ | $e^{i\pi t_2}\sigma_0$ | $-i\sigma_y$ | $\sigma_z$ | $\sigma_x$ | $a+b$ |
| $C_1$ | $e^{i\pi(t_1+t_2)}\sigma_0$ | $-i\sigma_y$ | $\sigma_z$ | $\sigma_x$ | $a+b$ |
| $B_1$ | $e^{i\pi t_3}\sigma_0$ | $\sigma_x$ | $\sigma_z$ | $-i\sigma_y$ | $a+b$ |
| $A_1$ | $e^{i\pi(t_1+t_3)}\sigma_0$ | $\sigma_x$ | $\sigma_z$ | $-i\sigma_y$ | $a+b$ |
| $D_1^+$ | $e^{i\pi(t_2+t_3)}$ | $i$ | 1 | $i$ | $a+y$ |
| $D_1^-$ | $e^{i\pi(t_2+t_3)}$ | $i$ | $-1$ | $-i$ | $b-y$ |
| $D_2^+$ | $e^{i\pi(t_2+t_3)}$ | $-i$ | 1 | $-i$ | $a+y$ |
| $D_2^-$ | $e^{i\pi(t_2+t_3)}$ | $-i$ | $-1$ | $i$ | $b-y$ |
| $E_1^+$ | $e^{i\pi(t_1+t_2+t_3)}$ | $i$ | 1 | $i$ | $a+z$ |
| $E_1^-$ | $e^{i\pi(t_1+t_2+t_3)}$ | $i$ | $-1$ | $-i$ | $b-z$ |
| $E_2^+$ | $e^{i\pi(t_1+t_2+t_3)}$ | $-i$ | 1 | $-i$ | $a+z$ |
| $E_2^-$ | $e^{i\pi(t_1+t_2+t_3)}$ | $-i$ | $-1$ | $i$ | $b-z$ |

**Table 4.4** Primitive lattice and reciprocal lattice vectors in monoclinic primitive and base-centered lattices. Here we set the lattice constants to be unity for simplicity

| Lattice | Space groups | $\mathbf{a}_1, \mathbf{a}_2, \mathbf{a}_3$ | $\mathbf{b}_1, \mathbf{b}_2, \mathbf{b}_3$ |
|---|---|---|---|
| Monoclinic primitive | ***7***, ***13***, ***14***, etc. | (100), (010), (001) | $2\pi(100), 2\pi(010), 2\pi(001)$ |
| Monoclinic base-centered | ***9***, ***15***, etc. | $(\frac{1}{2}\frac{1}{2}0), (\frac{\bar{1}}{2}\frac{1}{2}0)$, (001) | $2\pi(110), 2\pi(\bar{1}10), 2\pi(001)$ |

reciprocal primitive vectors, the $k_y = \pi$ plane is not glide-invariant locally. Therefore, it is necessary to modify the formula of the $Z_2$ topological invariant associated with the glide-symmetric systems.

In ***7***, every **k** point on the $k_y = \pi$ plane is invariant under the glide operation,

$$\mathbf{k} = \hat{G}_y\mathbf{k} \pmod{\mathbf{b}_i^{Pc}}. \tag{4.47}$$

where $\mathbf{b}_i^{Pc}$ $(i = 1, 2, 3)$ are the reciprocal lattice vectors for ***7*** $(Pc)$ given in Table 4.4. Therefore, the Bloch Hamiltonian on the $k_y = \pi$ plane can be block-diagonalized into two blocks characterized by glide eigenvalues, and eigenstates on the $k_y = \pi$ plane are classified with respect to the glide sectors. On the other hand, **k** points on the $k_y = \pi$ plane are not invariant under the glide operation in ***9*** because $(0, 2\pi, 0)$ is not among the reciprocal vectors for ***9*** $(Cc)$ given by Table 4.4. Therefore, eigenstates on $k_y = \pi$ cannot be associated with glide sectors. Therefore, we need to alter the formula of the glide-$Z_2$ invariant for ***9*** due to this property related to $k_y = \pi$ plane. This is due to the nonprimitive nature of ***9***.

Our strategy is the following. To construct a formula of the glide-$Z_2$ invariant for ***9***, we will double the unit cell of ***9*** into that of ***7***, leading to a folding of the BZ of ***9*** into that of ***7***. Note that both the $k_y = \pi$ plane and $k_y = -\pi$ plane in ***9*** are projected into the $k_y = \pi$ plane in ***7*** by folding the BZ. While the two points $(k_x, \pi, k_z)$ and $(k_x, -\pi, k_z)$ are distinct in ***9***, they are projected into the same point **k** in the BZ of ***7***. Thus, by multiplying the unit cell, we can connect the wavefunctions between ***7*** and ***9***, and we address how the terms on the $k_y = \pi$ plane are described in ***9***.

Henceforth, $O$ denotes an operator operating on or a function calculated from wave functions $|u_{n\mathbf{k}}\rangle$ in ***9***, whereas $\tilde{O}$ denotes an operator operating on or a function calculated from wave functions $|\tilde{u}_{n\mathbf{k}}\rangle$ in ***7***, respectively, throughout the present section.

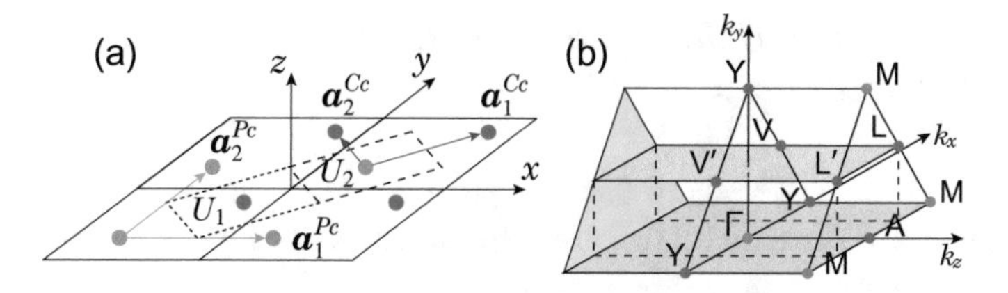


**Fig. 4.3** **a** Illustration of the relation between the primitive lattice in ***7*** and the base-centered lattice in ***9***. $U_1$ and $U_2$ denote two unit cells of the base-centered lattice ***9*** and $U_1 + U_2$ is a unit cell of the primitive lattice ***7***. **b** Upper half of the Brillouin zone of the monoclinic base-centered lattice in ***9*** ($Cc$) and ***15*** ($C2/c$). $\Gamma$, $Y$, $V(V')$, $A$, $M$, and $L(L')$ denote the high-symmetry points in ***15***. The corresponding Brillouin zone of the monoclinic primitive lattice is depicted by dotted lines

## 4.4.2 Relation Between Hamiltonians in 9 and 7

We start by constructing the Hamiltonian for ***7*** from that of ***9***. As depicted in Fig. 4.3a, the unit cell of ***7*** consists of two unit cells $U_1$ and $U_2$ of ***9***, which are related by translation by a vector $\mathbf{a}_1 = \mathbf{a}_1^{Cc}$. Therefore, we can divide the Bloch Hamiltonian for ***9*** into two parts:

$$\mathcal{H}(\mathbf{k}) = \mathcal{H}_1(\mathbf{k}) + \mathcal{H}_2(\mathbf{k})e^{i\mathbf{k}\cdot\mathbf{a}_1}. \tag{4.48}$$

The Hamiltonian $\mathcal{H}_1(\mathbf{k})$ denotes intra-unit-cell hopping from $U_\alpha$ to $U_\alpha$ itself where $\alpha = 1, 2$, while $\mathcal{H}_2(\mathbf{k})$ denotes inter-unit-cell hopping between $U_1$ and $U_2$. Due to dividing of the unit cell, $\mathcal{H}_1(\mathbf{k})$ and $\mathcal{H}_2(\mathbf{k})$ satisfy the periodicity for ***7***, i.e., $\mathcal{H}_\alpha(\mathbf{k} + \mathbf{b}_i^{Pc}) = \mathcal{H}_\alpha(\mathbf{k})$ ($i = 1, 2, 3, \alpha = 1, 2$). Then, a set of the Hamiltonian and the glide operator for ***7*** is given by

$$\tilde{\mathcal{H}}(\mathbf{k}) = \begin{pmatrix} \mathcal{H}_1(\mathbf{k}) & \mathcal{H}_2(\mathbf{k})e^{-2i\mathbf{k}\cdot\mathbf{a}_1} \\ \mathcal{H}_2(\mathbf{k}) & \mathcal{H}_1(\mathbf{k}) \end{pmatrix}, \tag{4.49}$$

$$\tilde{G}_y(\mathbf{k}) = \begin{pmatrix} G_y(\mathbf{k}) & \\ & G_y(\mathbf{k})e^{-ik_y} \end{pmatrix}, \tag{4.50}$$

where $G_y(\mathbf{k})$ is the glide operator for ***9*** satisfying Eq. (4.2). Therefore, we can easily see

$$\tilde{G}_y(\mathbf{k})\tilde{\mathcal{H}}(\mathbf{k})\tilde{G}_y^{-1}(\mathbf{k}) = \tilde{\mathcal{H}}(g_y\mathbf{k}), \tag{4.51}$$

where $g_y\mathbf{k}$ is transformed from $\mathbf{k}$ by $\hat{G}_y$. Thus we have established the Hamiltonian and the glide operator for ***7*** out of those for ***9***.

### 4.4.3 Derivation of the Formula of the $Z_2$ Topological Invariant for *9*

In order to derive a formula of the $Z_2$ topological invariant for ***9***, we write down wavefunctions in ***7*** in terms of those in ***9***. In particular, on the $k_y = \pi$ plane, we need wavefunctions in each of the glide sectors.

Let $|u(\mathbf{k})\rangle$ denote the wavefunction in ***9*** with the energy $E_{\mathbf{k}}$: $\mathcal{H}(\mathbf{k})|u(\mathbf{k})\rangle = E_{\mathbf{k}}|u(\mathbf{k})\rangle$. Then, the wavefunction for ***7*** is given by

$$|\tilde{u}(\mathbf{k})\rangle = \frac{1}{\sqrt{2}}\begin{pmatrix} 1 \\ e^{i\mathbf{k}\cdot\mathbf{a}_1} \end{pmatrix}|u(\mathbf{k})\rangle, \tag{4.52}$$

and one can simply see that $\tilde{\mathcal{H}}(\mathbf{k})|\tilde{u}(\mathbf{k})\rangle = E_{\mathbf{k}}|\tilde{u}(\mathbf{k})\rangle$.

Now we construct wavefunctions in each glide sector in ***7*** on the $k_y = \pi$ plane. The glide symmetry guarantees

$$G_y(\mathbf{k})|u(\mathbf{k})\rangle = e^{i\chi(\mathbf{k})}|u(g_y\mathbf{k})\rangle, \tag{4.53}$$

where $\chi(\mathbf{k})$ is a real function. By applying $G_y(g_y\mathbf{k})$, we get

$$e^{-ik_z} = e^{i\chi(\mathbf{k})+i\chi(g_y\mathbf{k})}. \tag{4.54}$$

In ***9***, $|u(k_x,\pi,k_z)\rangle$ and $G_y(k_x,\pi,k_z)|u(k_x,\pi,k_z)\rangle \propto |u(k_x,-\pi,k_z)\rangle$ reside at different wavevectors. On the other hand, when we go to ***7***, they correspond to the same wavevector upon halving the BZ, and their linear combinations give wavefunctions within each glide sector. By applying $\tilde{G}_y(\mathbf{k})$ onto $|\tilde{u}(\mathbf{k})\rangle$ where $\mathbf{k} = (k_x,\pi,k_z)$, we have

$$\tilde{G}_y(\mathbf{k})|\tilde{u}(k_x,\pi,k_z)\rangle = \frac{1}{\sqrt{2}}\begin{pmatrix} 1 \\ -e^{-i\mathbf{k}\cdot\mathbf{a}_1} \end{pmatrix} e^{i\chi(k_x,\pi,k_z)}|u(k_x,-\pi,k_z)\rangle. \tag{4.55}$$

Thus, from the degenerate eigenstates $|\tilde{u}(\mathbf{k})\rangle$ and $\tilde{G}_y(\mathbf{k})|\tilde{u}(\mathbf{k})\rangle$, we can establish eigenstates in $\tilde{G}_y(\mathbf{k}) = \pm\exp(-ik_z/2)$ sectors by linear combinations:

$$\begin{aligned}|\tilde{u}^{\pm}(k_x,\pi,k_z)\rangle = \frac{1}{2}\left[\begin{pmatrix} 1 \\ ie^{ik_z/2} \end{pmatrix}|u(k_x,\pi,k_z)\rangle + \begin{pmatrix} 1 \\ -ie^{ik_z/2} \end{pmatrix} e^{i\chi(k_x,\pi,k_z)+ik_z/2}|u(k_x,-\pi,k_z)\rangle\right] \\ \times\exp\left[-i\left(\chi(\pi,\pi,k_z)+\frac{k_z}{2}\right)\frac{k_x}{2\pi}\right]\times\begin{cases} 1 & (g_+\ \text{sector}) \\ e^{ik_x/2} & (g_-\ \text{sector}) \end{cases}.\end{aligned} \tag{4.56}$$

The phase factor $e^{-i(\chi(\pi,\pi,k_z)+k_z/2)\frac{k_x}{2\pi}}$ ($e^{-i(\chi(\pi,\pi,k_z)+k_z/2)\frac{k_x}{2\pi}}e^{ik_x/2}$) is required by imposing the periodicity between $k_x = \pi$ and $k_x = -\pi$ in ***7***.

We plug in these eigenstates into the formula of the glide $Z_2$ invariant of ***7*** in Eq. (4.12). In particular, we can recast the integral of the Berry curvature for the $g_+$

sector within the $k_y = \pi$ plane in $\mathbf{\mathit{9}}$ into

$$\begin{aligned}
&\int_{-\pi}^{\pi} dk_z \int_{-\pi}^{\pi} dk_x \mathrm{tr} \tilde{\mathcal{F}}_{zx}^{+}(k_x, \pi, k_z) \\
&= \frac{1}{2}\left( \int_{-\pi}^{\pi} dk_z \int_{-\pi}^{\pi} dk_x \mathrm{tr} \mathcal{F}_{zx}(k_x, \pi, k_z) + \int_{-\pi}^{\pi} dk_z \int_{-\pi}^{\pi} dk_x \mathrm{tr} \mathcal{F}_{zx}(k_x, -\pi, k_z) \right) \\
&\quad + \frac{1}{4}\int_{-\pi}^{\pi} dk_x \mathrm{tr}\left( i \langle u(k_x, \pi, -\pi)|G(k_x, \pi, -\pi)|u(k_x, \pi, -\pi)\rangle + \text{c.c.} \right). \qquad (4.57)
\end{aligned}$$

The Berry phase term on the $k_y = \pi$ plane is also expressed as

$$\begin{aligned}
&2\oint_{-\pi}^{\pi} dk_x \mathrm{tr} \tilde{\mathcal{A}}_x^{-}(k_x, \pi, -\pi) \\
&= \int_{-\pi}^{\pi} dk_x \mathrm{tr} \mathcal{A}_x(k_x, \pi, -\pi) \\
&\quad + \int_{-\pi}^{\pi} dk_x \mathrm{tr} \mathcal{A}_x(k_x, -\pi, -\pi) + \chi(\pi, \pi, -\pi) + \chi(-\pi, \pi, -\pi) \\
&\quad + \frac{1}{4}\int_{-\pi}^{\pi} dk_x \mathrm{tr}\left( i \langle u(k_x, \pi, -\pi)|G(k_x, \pi, -\pi)|u(k_x, \pi, -\pi)\rangle + \text{c.c.} \right). \qquad (4.58)
\end{aligned}$$

The details of calculations for $\tilde{\mathcal{A}}$ and $\tilde{\mathcal{F}}$ are given in Sect. 4.7.

Next, we consider the $k_z = -\pi$ plane. The term on this plane for $\mathbf{\mathit{9}}$ is rewritten as

$$\begin{aligned}
&\int_{0}^{\pi} dk_y \int_{-\pi}^{\pi} dk_x \mathrm{tr} \tilde{\mathcal{F}}_{xy}(k_x, k_y, -\pi) \\
&= \int_{0}^{2\pi} dk_y \int_{k_y - 2\pi}^{-k_y + 2\pi} dk_x \mathrm{tr} \mathcal{F}_{xy}(k_x, k_y, -\pi) \\
&\quad + \int_{-\pi}^{\pi} dk\, \mathrm{tr} \mathcal{A}_k(k - \pi, k + \pi, -\pi) + \int_{-\pi}^{\pi} dk\, \mathrm{tr} \mathcal{A}_k(k + \pi, -k + \pi, -\pi) \\
&\quad + \int_{-\pi}^{\pi} dk_x \mathrm{tr} \mathcal{A}_x(k_x, \pi, -\pi) + \int_{-\pi}^{\pi} dk_x \mathrm{tr} \mathcal{A}_x(k_x, -\pi, -\pi), \qquad (4.59)
\end{aligned}$$

where $k$ is a parameter. The domain of integration on the left-hand side in Eq. (4.59) is depicted in Fig. 4.4a, while the domain and the loops of integration on the right-hand side in Eq. (4.59) are shown in Fig. 4.4b. We have used the periodicity of BZ of $\mathbf{\mathit{9}}$. We can proceed by noticing that the second and third terms in Eq. (4.59) are related by glide symmetry and the periodicity of BZ upon a gauge choice related to the gauge function $\chi(\mathbf{k})$. To make progress, we have to clarify the properties of the gauge function $\chi(\mathbf{k})$. The gauge function terms in Eq. (4.58) and the Berry phase terms in Eq. (4.59) are not gauge invariant. Nonetheless, the summation of those terms is gauge invariant. We will show this argument in the following.

At first, it is obvious from Eq. (4.54) that

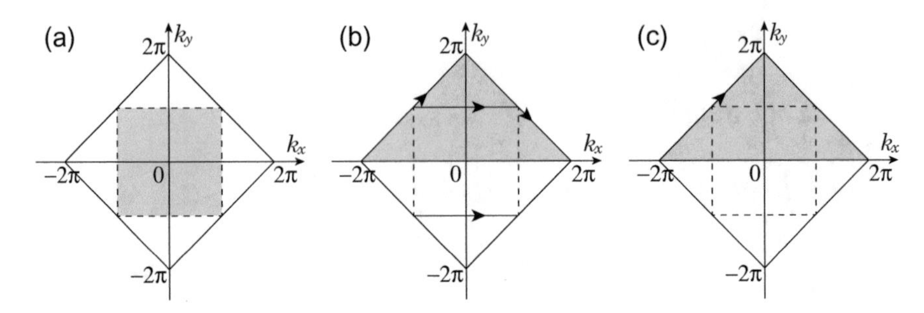


**Fig. 4.4** Domain of integration for rewriting the terms of Berry connections and curvatures. The dashed and solid lines indicate the Brillouin zones on the $k_z = -\pi$ plane in ***7*** and ***9***, respectively

$$\chi(\mathbf{k}) + \chi(g_y\mathbf{k}) = -k_z \pmod{2\pi}. \tag{4.60}$$

Owing to the periodicity of BZ, we can write

$$\chi(\mathbf{k} + \mathbf{b}_1^{Cc}) = \chi(\mathbf{k}) + 2\pi N, \tag{4.61}$$

$$\chi(\mathbf{k} - \mathbf{b}_2^{Cc}) = \chi(\mathbf{k}) + 2\pi N', \tag{4.62}$$

where $N$ and $N'$ are integers. By substituting $\mathbf{k} \to \mathbf{k} + \mathbf{b}_1^{Cc}$ in Eq. (4.60), we get $N = N'$. Then we obtain an expression of $\chi(\mathbf{k})$ as

$$\chi(\mathbf{k}) = Nk_y + \theta(k_x, k_y), \tag{4.63}$$

where $N$ is an integer and $\theta(k_x, k_y)$ is a periodic function of $k_x$ and $k_y$ in the BZ. For all of the occupied states hence

$$\sum_i (\chi_i(\pi, \pi, -\pi) + \chi_i(-\pi, \pi, -\pi)) = \sum_i 2\pi N_i \pmod{4\pi}, \tag{4.64}$$

in which $N_i$ is an integer for the $i$-th occupied state. Secondly, we investigate the Berry phase terms in Eq. (4.59). Due to the periodicity of the BZ, we have

$$\int_{-\pi}^{\pi} dk\, \mathrm{tr}\mathcal{A}_k(k + \pi, -k + \pi, -\pi) = -\sum_i 2\pi N_i + \int_{-\pi}^{\pi} dk\, \mathrm{tr}\mathcal{A}_k(k - \pi, k + \pi, -\pi), \tag{4.65}$$

where the values of $N_i$ depend on the gauge choice. Hence, we conclude that the summation of $\chi(\mathbf{k})$ terms in Eq. (4.58) and the Berry phase terms in Eq. (4.59) is gauge invariant. Therefore, the first term and two times the second term on the right-hand side in Eq. (4.59) only contribute to the glide-$Z_2$ invariant, and its domain and loop of integration are shown in Fig. 4.4c.

The terms on the glide-invariant plane $k_y = 0$ in ***9*** are given by

$$\int_{-\pi}^{\pi} dk_z \int_{-\pi}^{\pi} dk_x \mathrm{tr}\tilde{\mathcal{F}}_{zx}^{-}(k_x, 0, k_z) = \int_{-\pi}^{\pi} dk_z \int_{-2\pi}^{2\pi} dk_x \mathrm{tr}\mathcal{F}_{zx}^{-}(k_x, 0, k_z), \tag{4.66}$$

and

$$\oint_{-\pi}^{\pi} dk_x \mathrm{tr}\tilde{\mathcal{A}}_x^{+}(k_x, 0, -\pi) = \int_{-2\pi}^{2\pi} dk_x \mathrm{tr}\mathcal{A}_x^{+}(k_x, 0, -\pi). \tag{4.67}$$

Therefore, by combining these results, the glide-$Z_2$ invariant for ***9*** recast as follows:

$$\begin{aligned}\tilde{\nu} = \frac{1}{2\pi}\Bigg[ & 2\int_{-\pi}^{\pi} dk\, \mathrm{tr}\mathcal{A}_k(k-\pi, k+\pi, -\pi) - 2\int_{-2\pi}^{2\pi} dk_x \mathrm{tr}\mathcal{A}_x^{+}(k_x, 0, -\pi) \\ & + \int_0^{2\pi} dk_y \int_{k_y-2\pi}^{-k_y+2\pi} dk_x \mathrm{tr}\mathcal{F}_{xy}(k_x, k_y, -\pi) + \int_{-\pi}^{\pi} dk_z \int_{-2\pi}^{2\pi} dk_x \mathrm{tr}\mathcal{F}_{zx}^{-}(k_x, 0, k_z) \\ & - \frac{1}{2}\left( \int_{-\pi}^{\pi} dk_z \int_{-\pi}^{\pi} dk_x \mathrm{tr}\mathcal{F}_{zx}(k_x, \pi, k_z) + \int_{-\pi}^{\pi} dk_z \int_{-\pi}^{\pi} dk_x \mathrm{tr}\mathcal{F}_{zx}(k_x, -\pi, k_z) \right) \Bigg].\end{aligned} \tag{4.68}$$

Because we are assuming an insulating system, the Chern number on the $k_x$-$k_z$ plane is independent on the value of $k_y$. Noticing that the summation of the last two terms in Eq. (4.68) is nothing but the Chern number on the glide-invariant plane $k_y = 0$, we have an alternative expression for $\tilde{\nu}$,

$$\begin{aligned}\tilde{\nu} = \frac{1}{2\pi}\Bigg[ & 2\int_{-\pi}^{\pi} dk\, \mathrm{tr}\mathcal{A}_k(k-\pi, k+\pi, -\pi) - 2\int_{-2\pi}^{2\pi} dk_x \mathrm{tr}\mathcal{A}_x^{+}(k_x, 0, -\pi) \\ & + \int_0^{2\pi} dk_y \int_{k_y-2\pi}^{-k_y+2\pi} dk_x \mathrm{tr}\mathcal{F}_{xy}(k_x, k_y, -\pi) - \int_{-\pi}^{\pi} dk_z \int_{-2\pi}^{2\pi} dk_x \mathrm{tr}\mathcal{F}_{zx}^{+}(k_x, 0, k_z) \\ & + \frac{1}{2}\oint_{-\pi}^{\pi} dk_z \oint_{-2\pi}^{2\pi} dk_x \mathrm{tr}\mathcal{F}_{zx}(k_x, 0, k_z) \Bigg].\end{aligned} \tag{4.69}$$

We have used the relation $\mathrm{tr}\mathcal{F}_{ij}(\mathbf{k}) = \mathrm{tr}\mathcal{F}_{ij}^{+}(\mathbf{k}) + \mathrm{tr}\mathcal{F}_{ij}^{-}(\mathbf{k})$.

### 4.4.4 Topological Invariants for 15

Next, we consider what happens when inversion symmetry is added. The space group becomes ***15***. We derive the formula for ***15*** in two ways; one is to exploit a sewing matrix introduced in Ref. [9], the other is to exploit the compatibility relations in the band theory from the formula for ***13*** we showed in the previous section. As a result, we will show that the glide-$Z_2$ invariant for ***15*** is given by

$$(-1)^{\tilde{\nu}} = \prod_{i \in \mathrm{occ}} \zeta_i^+(\Gamma)\xi_i(V)\frac{\xi_i(Y)}{\zeta_i^+(Y)}, \tag{4.70}$$

where $\zeta_i^+$ is a $C_2$ eigenvalue in the $g_+$ sector and $\xi_i$ is an inversion parity for the $i$-th occupied state.

In addition, we find that the Chern number on the $k_x$-$k_z$ plane is also written in terms of the irreps as

$$C_y \in 2\mathbb{Z}, \quad (-1)^{C_y/2} = \prod_{i \in \mathrm{occ}} \frac{\xi_i(\mathrm{Y})\xi_i(\mathrm{V})}{\xi_i(\mathrm{M})\xi_i(\mathrm{L})}, \tag{4.71}$$

where $C_y$ denotes the Chern number on the $k_y = 0$ plane. The high-symmetry points are specified in Fig. 4.3b.

#### 4.4.4.1 Approach by Using Sewing Matrix

Let us start with symmetry consideration. Adding an inversion $\hat{I} = \{I|\mathbf{0}\}$ around the origin to ***9*** leads ***15***. Then $C_2$ rotational symmetry around the axis $x = 0, z = 1/4$,

$$\hat{C}_2 = \hat{G}_y\hat{I} = \left\{C_{2y}\middle|\frac{1}{2}\hat{\mathbf{z}}\right\}, \tag{4.72}$$

is also added to symmetry operations. The algebra is given by

$$\hat{C}_2\hat{G}_y = \hat{G}_y\hat{C}_2\{E|\hat{\mathbf{z}}\}. \tag{4.73}$$

We now derive the glide-$Z_2$ invariant for ***15*** with the help of the additional symmetries. We will calculate $(-1)^{\tilde{\nu}} = e^{i\pi\tilde{\nu}}$ with $\tilde{\nu}$ given by Eq. (4.69). First, due to the $C_2$ symmetry, the Berry curvature on the $k_z = -\pi$ plane satisfies

$$\mathrm{tr}\mathcal{F}_{xy}(k_x, k_y, -\pi) = -\mathrm{tr}\mathcal{F}_{xy}(-k_x, k_y, \pi) = -\mathrm{tr}\mathcal{F}_{xy}(-k_x, k_y, -\pi), \tag{4.74}$$

and hence we have

$$\int_0^{2\pi} dk_y \int_{k_y-2\pi}^{-k_y+2\pi} dk_x \mathrm{tr}\mathcal{F}_{xy}(k_x, k_y, -\pi) = 0. \tag{4.75}$$

Second, we move on to the Berry phase term $\int_{-\pi}^{\pi} dk\ \mathrm{tr}\mathcal{A}_k(k-\pi, k+\pi, -\pi)$. Because of inversion symmetry and the periodicity of the BZ,

$$\exp\left[i\int_{-\pi}^{\pi} dk\ \mathrm{tr}\mathcal{A}_k(k-\pi, k+\pi, -\pi)\right] = \prod_{i \in \mathrm{occ}} \frac{\xi_i(L)}{\xi_i(M)}, \tag{4.76}$$

in which $\xi_i(P)$ is the inversion parity of the $i$-th occupied state at a high-symmetry point $P$.

Next, we address the terms for the glide-invariant plane $k_y = 0$. Equation (4.73) implies that the glide sector is unchanged upon the $C_2$ rotation. Therefore, the Berry curvature $\mathcal{F}^{\pm}_{zx}(\mathbf{k})$ within this plane have to be an even function of $k_x$ and $k_z$,

$$\mathrm{tr}\mathcal{F}^{\pm}_{zx}(k_x, 0, k_z) = \mathrm{tr}\mathcal{F}^{\pm}_{zx}(-k_x, 0, -k_z). \tag{4.77}$$

By using the Stokes' theorem on the $k_y = 0$ plane, we obtain

$$\begin{aligned}
&\exp\left[-\frac{i}{2}\int_{-\pi}^{\pi} dk_z \int_{-2\pi}^{2\pi} dk_x \mathrm{tr}\mathcal{F}^{+}_{zx}(k_x, 0, k_z) - i\oint_{-2\pi}^{2\pi} dk_x \mathrm{tr}\mathcal{A}^{+}_{x}(k_x, 0, -\pi)\right] \\
&= \exp\left[i\int_{0}^{\pi} dk_z \int_{-2\pi}^{2\pi} dk_x \mathrm{tr}\mathcal{F}^{+}_{zx}(k_x, 0, k_z) - i\oint_{-2\pi}^{2\pi} dk_x \mathrm{tr}\mathcal{A}^{+}_{x}(k_x, 0, -\pi)\right] \\
&= \exp\left[-i\oint_{-2\pi}^{2\pi} dk_x \mathrm{tr}\mathcal{A}^{+}_{x}(k_x, 0, 0)\right] = \prod_{i\in\mathrm{occ}} \frac{\zeta_i^{+}(\Gamma)}{\zeta_i^{+}(Y)},
\end{aligned} \tag{4.78}$$

in which $\zeta_i(P)$ is the $C_2$ eigenvalue at a high-symmetry point $P$.

Finally, we study the half of the Chern number on the $k_y = 0$ plane. Since the flux of the Berry curvatures has to be preserved in an insulator, the Chern number on the $k_y = 0$ plane is equal to the sum of those on the planes perpendicular to $\mathbf{b}_1^{Cc}$ and $\mathbf{b}_2^{Cc}$, namely the integral of the Berry curvature along the planes $\{(k-\pi, k+\pi, k_z)| -\pi < k < \pi, -\pi < k_z < \pi\}$ (planes $\mathcal{D}_1$ and $\mathcal{D}_2$) and $\{(k+\pi, k-\pi, k_z)| -\pi < k < \pi, -\pi < k_z < \pi\}$ (planes $\mathcal{D}_3$ and $\mathcal{D}_4$) depicted in Fig. 4.5. Therefore, we have

$$\begin{aligned}
\frac{1}{2}\oint_{-\pi}^{\pi} dk_z \oint_{-2\pi}^{2\pi} dk_x \mathrm{tr}\mathcal{F}_{zx}(k_x, 0, k_z) &= \frac{1}{2}\int_{\mathcal{D}_1+\mathcal{D}_2+\mathcal{D}_3+\mathcal{D}_4} d^2k\ \mathrm{tr}\mathcal{F}(k-\pi, k+\pi, k_z) \\
&= 2\oint_{-\pi}^{\pi} dk_z \int_{-\pi}^{0} dk\ \mathrm{tr}\mathcal{F}(k-\pi, k+\pi, k_z),
\end{aligned} \tag{4.79}$$

where the domains of integration $\mathcal{D}_{1,2,3,4}$ are shown in Fig. 4.5, and we have used the fact that the integrals of the Berry curvature are all the same for $\mathcal{D}_i (i = 1, \dots, 4)$ due to the periodicity of the BZ, inversion symmetry, and $C_2$ rotational symmetry. By using the Stokes' theorem, it leads to

$$\begin{aligned}
&\exp\left[i\oint_{-\pi}^{\pi} dk_z \int_{-\pi}^{0} dk \mathrm{tr}\mathcal{F}(k-\pi, k+\pi, k_z)\right] \\
&= \exp\left[i\oint_{-\pi}^{\pi} dk_z \mathrm{tr}\mathcal{A}^{+}_{z}(-2\pi, 0, k_z) + i\oint_{-\pi}^{\pi} dk_z \mathrm{tr}\mathcal{A}^{+}_{z}(-\pi, \pi, k_z)\right] = \prod_{i\in\mathrm{occ}} \frac{\xi_i(Y)\xi_i(V)}{\xi_i(M)\xi_i(L)}
\end{aligned} \tag{4.80}$$

from Eq. (4.76).

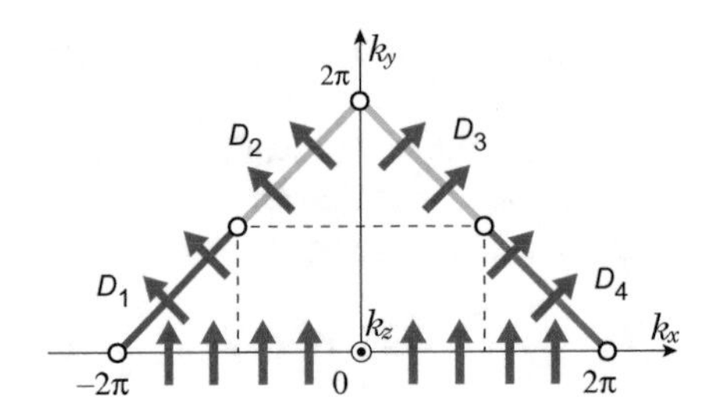


**Fig. 4.5** Schematic figure of the flux of the Berry curvatures on the arbitrary $k_z$ plane denoted by red arrows. The dashed and solid lines indicate the Brillouin zones in ***7*** and ***9***, respectively. $\mathcal{D}_1, \mathcal{D}_2, \mathcal{D}_3$, and $\mathcal{D}_4$ denote the planes $\mathcal{D}_1 : \{(k-\pi, k+\pi, k_z)| -\pi < k < 0, -\pi < k_z < \pi\}$, $\mathcal{D}_2 : \{(k-\pi, k+\pi, k_z)|0 < k < \pi, -\pi < k_z < \pi\}$, $\mathcal{D}_3 : \{(k+\pi, k-\pi, k_z)| -\pi < k < 0, -\pi < k_z < \pi\}$, and ,$\mathcal{D}_4 : \{(k+\pi, k-\pi, k_z)|0 < k < \pi, -\pi < k_z < \pi\}$, respectively. The flux of the Berry curvatures is preserved in an insulating system, and its contributions on the $\mathcal{D}_1, \mathcal{D}_2, \mathcal{D}_3$, and $\mathcal{D}_4$ are all the same due to the periodicity of the Brillouin zone in ***9***, inversion symmetry, and twofold rotational symmetry

As a consequence, in the presence of inversion symmetry, the glide-$Z_2$ invariant for ***15*** is rewritten as

$$(-1)^{\tilde{\nu}} = \prod_{i\in\text{occ}} \frac{\xi_i(L)}{\xi_i(M)} \frac{\zeta_i^+(\Gamma)}{\zeta_i^+(Y)} \frac{\xi_i(Y)\xi_i(V)}{\xi_i(M)\xi_i(L)}$$
$$= \prod_{i\in\text{occ}} \zeta_i^+(\Gamma)\xi_i(V)\frac{\xi_i(Y)}{\zeta_i^+(Y)}. \tag{4.81}$$

The states with the positive $C_2$ eigenvalue ($\zeta_i = +1$) or those with the even-parity ($\xi_i = +1$) do not contribute to the nontrivial glide-$Z_2$ invariant $\tilde{\nu}$ from Eq. (4.81). Thus, we have only to count the number of states giving nontrivial contribution, whose irreps are given by $\Gamma_2^-$ at $\Gamma$, $V_1^-$ at $V$, and $Y_1^-$ at $Y$, respectively. Those irreps at the high-symmetry points are summarized in Table 4.5. Therefore, we eventually have a formula of $\tilde{\nu}$ expressed in terms of irreps for ***15*** as

$$\nu = N_{\Gamma_2^-}(\Gamma) + N_{Y_1^-}(\mathrm{Y}) + N_{V_1^-}(\mathrm{V}), \tag{4.82}$$

where $N_R(P)$ is the number of occupied states at a high-symmetry point $P$ with an irrep $R$ in the $k$-group of ***15*** at $P$.

In this class of systems in ***15***, the symmetry-based indicators are given by $\mathbb{Z}_2 \times \mathbb{Z}_2$ in Ref. [6, 7]. One $\mathbb{Z}_2$ is given by $\tilde{\nu}$ in Eq. (4.82), and the other $\mathbb{Z}_2$ is in fact a half of the Chern number on the glide plane. When we define the Chern numbers $C_1$ and $C_2$ on the planes normal to $\mathbf{b}_1^{Cc}$ and $\mathbf{b}_2^{Cc}$, respectively, one always has $C_1 = C_2$ and $C_y = C_1 + C_2$ as we have shown in Eq. (4.79) (Fig. 4.5). Therefore, in the present case the Chern number $C_y$ on the glide plane is always an even integer. As we have shown in Eqs. (4.79) and (4.80), this expression is immediately obtained as

**Table 4.5** Summary of irreducible representations of the little group and the number of states with that irreducible representation for ***15*** where $a, b, m, x, y$, and $z$ are an integer

| Seitz symbols | $\{1\|t_1, t_2, t_3\}$ | $\{2_{010}\|0, 0, 1/2\}$ | $\{\bar{1}\|0, 0, 0\}$ | $\{m_{010}\|0, 0, 1/2\}$ | Number of states |
|---|---|---|---|---|---|
| Matrix presentation | $\begin{pmatrix} 1 & 0 & 0 & t_1 \\ 0 & 1 & 0 & t_2 \\ 0 & 0 & 1 & t_3 \end{pmatrix}$ | $\begin{pmatrix} -1 & 0 & 0 & 0 \\ 0 & 1 & 0 & 0 \\ 0 & 0 & -1 & 1/2 \end{pmatrix}$ | $\begin{pmatrix} -1 & 0 & 0 & 0 \\ 0 & -1 & 0 & 0 \\ 0 & 0 & -1 & 0 \end{pmatrix}$ | $\begin{pmatrix} 1 & 0 & 0 & 0 \\ 0 & -1 & 0 & 0 \\ 0 & 0 & 1 & 1/2 \end{pmatrix}$ | |
| $\Gamma_1^+$ | 1 | 1 | 1 | 1 | $a$ |
| $\Gamma_1^-$ | 1 | 1 | −1 | −1 | $b+m$ |
| $\Gamma_2^+$ | 1 | −1 | 1 | −1 | $a-m$ |
| $\Gamma_2^-$ | 1 | −1 | −1 | 1 | $b$ |
| $Y_1^+$ | $e^{i2\pi t_2}$ | 1 | 1 | 1 | $a+x$ |
| $Y_1^-$ | $e^{i2\pi t_2}$ | 1 | −1 | −1 | $b+m-x$ |
| $Y_2^+$ | $e^{i2\pi t_2}$ | −1 | 1 | −1 | $a-m+x$ |
| $Y_2^-$ | $e^{i2\pi t_2}$ | −1 | −1 | 1 | $b-x$ |
| $V_1^+$ | $e^{i\pi(t_1+t_2)}$ | | 1 | | $a+b+y$ |
| $V_1^-$ | $e^{i\pi(t_1+t_2)}$ | | −1 | | $a+b-y$ |
| $A_1$ | $e^{i\pi t_3}\sigma_0$ | $\sigma_x$ | $\sigma_z$ | $-i\sigma_y$ | $a+b$ |
| $M_1$ | $e^{i\pi(2t_2+t_3)}\sigma_0$ | $\sigma_x$ | $\sigma_z$ | $-i\sigma_y$ | $a+b$ |
| $L_1^+$ | $e^{i\pi(t_1+t_2+t_3)}$ | | 1 | | $a+b+z$ |
| $L_1^-$ | $e^{i\pi(t_1+t_2+t_3)}$ | | −1 | | $a+b-z$ |
| $\Lambda_1$ | $e^{i2\pi t_2 v}$ | 1 | | | $a+b+m$ |
| $\Lambda_1$ | $e^{i2\pi t_2 v}$ | −1 | | | $a+b-m$ |
| $B_1$ | $e^{i2\pi(t_1 u+t_3 w)}$ | | | $e^{i\pi w}$ | $a+b$ |
| $B_2$ | $e^{i2\pi(t_1 u+t_3 w)}$ | | | $e^{i\pi(1+w)}$ | $a+b$ |

$$C_1 = C_2 = \frac{C_y}{2}, \quad C_y \in 2\mathbb{Z}, \quad (-1)^{C_y/2} = \prod_{i \in \mathrm{occ}} \frac{\xi_i(Y)\xi_i(V)}{\xi_i(M)\xi_i(L)}. \tag{4.83}$$

Because the product of the inversion parities at $A$, $M$, $Y$, and $V$ on the $k_y$ planes is always $+1$ thanks to the compatibility relation, the Chern number $C_y$ on the $k_y = 0$ plane should be even.

#### 4.4.4.2 Compatibility Relation from *13*

Here we show that Eq. (4.82) is compatible with the glide-$Z_2$ invariant in ***13*** we derived in Eq. (4.29),

$$\nu = N_{\Gamma_2^+}(\Gamma) + N_{Y_2^+}(Y) + N_{Z_2^-}(Z) + N_{C_2^-}(C), \tag{4.84}$$

in which $N_R(P)$ is the number of occupied states with an irreps $R$ for the $k$-group of ***13*** at a high-symmetry point $P$ that is given by Table 4.2.

Systems in ***15*** can be regarded as those in ***13*** by choosing the primitive lattice vectors to be the lattice vectors for ***13***, $\mathbf{a}_1 = (1, 0, 0)$, $\mathbf{a}_2 = (0, 1, 0)$, $\mathbf{a}_3 = (0, 0, 1)$. Thereby, the unit cell is doubled, and the BZ is folded by half, as we discussed in the previous subsection. On the $k_y = 0$ plane, the high-symmetry points $\Gamma_{15}$ and $Y_{15}$ in ***15*** are projected onto $\Gamma_{13}$ in ***13***, and $V_{15}$ is projected onto $C_{13}$.[1] Similarly, on the $k_y = \pi$ plane, $A_{15}$ and $M_{15}$ are projected onto $B_{13}$, and $L_{15}$ is projected onto $E_{13}$, respectively. Hence, the irreps for $\Gamma_{15}$ and $Y_{15}$ ($A_{15}$ and $M_{15}$) in ***15*** should be directly compatible with those for $\Gamma_{13}$ ($B_{13}$) in ***13***. On the other hand, the case for $V_{15}$ ($L_{15}$) is not simple because two inequivalent points $V_{15}$ and $V'_{15}$ ($L_{15}$ and $L'_{15}$) in ***15*** (Fig. 4.3) are mapped onto the same point in ***13***.

Let us consider the state with $V_1^+$ at $V_{15}$ point $|u_{V_1^+}(V_{15})\rangle$: $I|u_{V_1^+}(V_{15})\rangle = +|u_{V_1^+}(V_{15})\rangle$. By introducing $|u_{V_R}(V'_{15})\rangle \equiv G_y|u_{V_1^+}(V_{15})\rangle$, we have

$$\begin{aligned} I|u_{V_R}(V'_{15})\rangle = IG_y|u_{V_1^+}(V_{15})\rangle = G_yI|u_{V_1^+}(V_{15})\rangle = G_y|u_{V_1^+}(V_{15})\rangle \\ = |u_{V_R}(V'_{15})\rangle. \end{aligned} \tag{4.85}$$

Thus, the corresponding state at $V'_{15}$ is also characterized by $V_1^+$. Note that $V_{15}$ and $V'_{15}$ in ***15*** are simultaneously projected onto $C_{13}$ in ***13*** in the halved BZ. Therefore, the states with $V_1^+(V_1^-)$ at $V_{15}$ in ***15*** corresponds to the irreps $C_1^+ + C_2^+$ ($C_1^- + C_2^-$) at $C_{13}$ in ***13***.

In the same manner, we can find similar relation between $L_{15}$ in ***15*** and $E_{13}$ in ***13***. We consider a state $|u_{L_1^+}(L_{15})\rangle$ obeying the $L_1^+$ irrep at $L_{15}$: $I|u_{L_1^+}(L_{15})\rangle = +|u_{L_1^+}(L_{15})\rangle$. By introducing $|u_{L_R}(L_{15})\rangle \equiv G_y|u_{L_1^+}(L_{15})\rangle$, it follows

$$I|u_{L_R}(L'_{15})\rangle = -G_y|u_{L_1^+}(L_{15})\rangle = -|u_{L_R}(L'_{15})\rangle. \tag{4.86}$$

[1] Compare Fig. 4.3 with Fig. 4.1.

Therefore, when the state at $L_{15}$ is in the $L_1^+$ irrep, the corresponding state at $L'_{15}$ is characterized by $L_1^-$, and vice versa. As a result by folding the BZ by half, the states at $L_{15}$ in ***15*** are projected onto those with the 2D irrep $E_1$ at $E_{13}$ in ***13***.

Next, we investigate how the irreps at $Y_{13}$ and $Z_{13}$ in ***13*** correspond to those in ***15***. Let us start with $Y_{13}$ in ***13***. This point splits into two points $B_{15} = (\pm\pi, 0, 0)$ in ***15***, which are invariant under the glide operation but not under inversion and $C_2$ rotation. If the state at $B_{15}$ is in the $g_+(g_-)$ sector, it is characterized by $B_1(B_2)$. In addition, the states with $B_1(B_2)$ at $B_{15}$ should be compatible with those with $Y_1^+ + Y_2^-$ $(Y_1^- + Y_2^+)$ at $Y_{15}$ from compatibility relations. Therefore, we find

$$N_{Y_2^+}(Y_{13}) = N_{B_2}(B_{15}) = N_{Y_1^-}(Y_{15}) + N_{Y_2^+}(Y_{15}). \tag{4.87}$$

Similarly, we obtain

$$N_{Z_2^-}(Z_{13}) = N_{\Lambda_2}(\Lambda_{15}) = N_{Y_2^+}(Y_{15}) + N_{Y_2^-}(Y_{15}). \tag{4.88}$$

We have used the fact that $Z_{13}$ in ***13*** is projected onto $\Lambda_{15} = (0, \pm\pi, 0)$ in ***15*** ensured by $C_2$ rotation but not by inversion symmetry and glide symmetry.

In summary, we express the glide-$Z_2$ invariant in ***13*** in terms of irreps for ***15*** from the compatibility relations,

$$\begin{aligned}\tilde{\nu} &= N_{\Gamma_2^+}(\Gamma_{13}) + N_{Y_2^+}(Y_{13}) + N_{Z_2^-}(Z_{13}) + N_{C_2^+}(C_{13}) \pmod 2 \\ &= N_{\Gamma_2^+}(\Gamma_{15}) + N_{Y_1^-}(Y_{15}) + N_{Y_2^+}(Y_{15}) + N_{Y_2^-}(Y_{15}) + N_{V_1^-}(V_{15}) \pmod 2 \\ &= N_{\Gamma_2^-}(\Gamma_{15}) + N_{Y_1^-}(\mathrm{Y}_{15}) + N_{V_1^-}(\mathrm{V}_{15}) \pmod 2. \end{aligned} \tag{4.89}$$

We have used the compatibility conditions in the last equality. This result is indeed consistent with Eq. (4.81).

## 4.5 Symmetry-Based Indicators and $K$-Theory

In this section, we compare our results with the previous works, the symmetry-based indicator theory [6, 7] and the classification based on the $K$-theory [5].

Let us begin with ***13***. The symmetry-based indicator for ***13*** is $\mathbb{Z}_2 \times \mathbb{Z}_2$ [6, 7]. Following Refs. [6, 7], we can show that one $\mathbb{Z}_2$ means the $Z_2$ topological invariant $\tilde{\nu}$ for the glide-symmetric systems in Eq. (4.28), and that the other $\mathbb{Z}_2$ is the Chern number $C_y$ modulo 2 in Eq. (4.32). Thus, when inversion symmetry is present, $\tilde{\nu}$ and $C_y$ (mod 2) are both symmetry-based indicators, expressed in terms of irreps at high-symmetry points.

Let us discuss the relationship between the higher-order TI and these topological invariants. In a system with only inversion symmetry, obeying ***2*** ($P\bar{1}$), symmetry-based indicators are given by $(\mathbb{Z}_2)^3 \times \mathbb{Z}_4$ [6, 7], where the three factors of $\mathbb{Z}_2$ are weak indices for Chern insulators, while the factor of $\mathbb{Z}_4$ corresponds to a strong index $z_4$ in Eq. (4.30), counting the number of eigenstates below the Fermi energy having an odd parity. When $z_4 = 1$ or $z_4 = 3$, it means that Weyl nodes appear in momentum space. The case with $z_4 = 2$ corresponds to a higher-order TI, leading to existence of 1D chiral hinge modes on the surface. In the present case of ***13***, Eq. (4.31) shows that the $z_4 = 2$ phase is equivalent to $\tilde{\nu} = 1 \pmod 2$. Thus, the system with nontrivial glide-$Z_2$ invariant is in the phase with $z_4 = 2$, and it is a higher-order TI ensured by inversion symmetry. Thus they exhibit 1D chiral hinge modes.

On the other hand, in the classification based on the $K$-theory [5], the topological invariant given by the $E_\infty^{2,0}$ term expressing a topological invariant in terms of integrals of wavefunctions is the Chern number $\mathbb{Z}$ in ***13*** (Table 4.1). We conclude that the topological phase characterized by the glide-$Z_2$ invariant is related to the higher-order TI with trivial weak indices $(\mathbb{Z}_2, \mathbb{Z}_2, \mathbb{Z}_2; \mathbb{Z}_4) = (0, 0, 0; 2)$, while that by an odd value of the Chern number is $(\mathbb{Z}_2, \mathbb{Z}_2, \mathbb{Z}_2; \mathbb{Z}_4) = (0, 1, 0; 0)$, where three $\mathbb{Z}_2$'s refer to the Chern numbers modulo 2 on the $k_x = 0$, $k_y = 0$, and $k_z = 0$ planes, and $\mathbb{Z}_4$ refers to the $z_4$ indicator in Eq. (4.30).

In ***14***, we have obtained a formula for the combination of the glide-$Z_2$ invariant $\tilde{\nu}$ and the Chern number $C_y$ expressed in terms of irreps, and this totally agrees with the symmetry-based indicator $\mathbb{Z}_2$ for ***14*** [6, 7]. Thus, one cannot uniquely determine the values of these topological invariants, solely from the $\mathbb{Z}_2$ symmetry-based indicator. Furthermore, both the glide-$Z_2$ topological phase and a Chern insulator with the Chern number equal to $4n + 2$ ($n$: integer) correspond to higher-order TIs with $z_4 = 2$, ensured by inversion symmetry. It corresponds to the $(\mathbb{Z}_2, \mathbb{Z}_2, \mathbb{Z}_2; \mathbb{Z}_4) = (0, 0, 0; 2)$ phase. On the other hand, the $E_\infty^{2,0}$ term of the $K$-theory classification [5] shows the $\mathbb{Z}$ topological invariant, i.e., the Chern number (Table 4.1).

In ***15***, topological classification is very similar to the case of ***13*** because the symmetry-based indicator for ***15*** is $\mathbb{Z}_2 \times \mathbb{Z}_2$ [6, 7] and the $E_\infty^{2,0}$ term of the $K$-theory classification [5] is the $\mathbb{Z}$ (Table 4.1). First, we show that the glide-$Z_2$ invariant in Eq. (4.82) is directly related to the $z_4$ indicator for centrosymmetric systems similar to Eq. (4.31):

$$\tilde{\nu} \equiv \frac{z_4}{2} \pmod 2. \tag{4.90}$$

In a similar manner with Sect. 4.3, we exploit the compatibility relations. The numbers of irreps $R$ at high-symmetry points are summarized in the rightmost column of Table 4.5. The integer parameters $a, b, \cdots$ correspond to the generators of the abelian group generated by the irreps at high-symmetry points. Therefore its dimension is given by $d_{\mathrm{BS}} = 6$ for ***15*** [6]. By using this, the glide-$Z_2$ invariant in Eq. (4.82) is equal to $a + b + m - x - y \pmod 2$. On the other hand, the $\mathbb{Z}_4$ index $z_4$ for centrosymmetric systems is equal to $z_4 = 2(a + b + m - x - y) = 2\tilde{\nu} \pmod 4$, and the glide-$Z_2$ invariant in Eq. (4.82) is exactly equivalent to the half of $z_4$ modulo 2. Therefore, we can conclude that one $\mathbb{Z}_2$ in the symmetry-based indicator is the glide-$Z_2$ topological invariant in Eq. (4.82), and that the other $\mathbb{Z}_2$ is the Chern number modulo 2 we have shown in Eq. (4.80). However, the $E_\infty^{2,0}$ term is $\mathbb{Z}$. This implies that the topological phase with a nontrivial glide-$Z_2$ invariant should be related to the higher-order TI with trivial weak indices $(\mathbb{Z}_2, \mathbb{Z}_2, \mathbb{Z}_2; \mathbb{Z}_4) = (0, 0, 0; 2)$, while that the Chern number on the glide plane equal to $4n+2$ ($n$: integer) is $(\mathbb{Z}_2, \mathbb{Z}_2, \mathbb{Z}_2; \mathbb{Z}_4) = (1, 1, 0; 0)$, where three $\mathbb{Z}_2$'s refer to the Chern numbers modulo 2 along $\mathbf{b}_i^{Cc}$ ($i = 1, 2, 3$). In fact, this conclusion is obvious because we have verified the relations between $\tilde{\nu}$ in ***13*** and that in ***15*** in the previous section. Therefore, our results totally agree with previous related works. This is among the principle results of the present chapter, and we check this scenario by constructing simple tight-binding models exhibiting topological phases in the following chapter.

## 4.6 Spinful Systems

We dealt with the spinless systems in the previous sections. Here we briefly mention the case of the spinful systems. The formulas for ***13*** in Eq. (4.28), ***14*** in Eq. (4.44), and ***15*** in Eq. (4.81) can be also applied in the same manner to the spinful systems, because the commutation relations between the glide and $C_2$ rotation/screw operations in Eqs. (4.21), (4.35), and (4.73) are unchanged, even though symmetry representations become double-valued. Consequently, the formulas for spinful systems of ***13***, ***14***, and ***15*** are given by

$$(-1)^{\tilde{\nu}} = \prod_{i \in \text{occ}} \frac{\zeta_i^-(\Gamma)\zeta_i^+(C)}{\zeta_i^-(Y)\zeta_i^+(Z)} \quad (\text{in } \textbf{\textit{13}}), \tag{4.91}$$

$$C_y \in 2\mathbb{Z}, \quad (-1)^{\tilde{\nu}}(-1)^{C_y/2} = \prod_{i \in \text{occ}} \frac{\eta_i^-(\Gamma)\eta_i^+(D)}{\eta_i^-(Y)\eta_i^+(E)} \quad (\text{in } \textbf{\textit{14}}), \tag{4.92}$$

$$(-1)^{\tilde{\nu}} = \prod_{i \in \text{occ}} \zeta_i^+(\Gamma)\xi_i(V)\frac{\xi_i(Y)}{\zeta_i^+(Y)} \quad (\text{in } \textbf{\textit{15}}), \tag{4.93}$$

where $\zeta_i^\pm = \pm i$ and $\eta_i^\pm = \pm i e^{-ik_y/2}$ are eigenvalues of the $C_2$ rotation/screw for the eigenstates in the $g_\pm$ sector,

$$g_\pm(k_z) = \pm i e^{-ik_z/2}, \tag{4.94}$$

at high-symmetry points $\Gamma$, $Y$, $Z$, and $C$ in ***13***, $\Gamma$, $Y$, $D$, and $E$ in ***14***, and $\Gamma$, $Y$, and $V$ in ***15***, respectively. Obviously, these formulas can be expressed in terms of irreps for double-valued groups.

## 4.7 Conclusion and Discussion

We studied the fate of the glide-symmetric $Z_2$ topological invariant when inversion symmetry is added while time-reversal symmetry is not enforced in this chapter. In ***13*** and ***15***, we derive the formulas for the glide-symmetric $Z_2$ topological invariant, and find that they are expressed in terms of the irreducible representations at high-symmetry points. The symmetry-based indicators are $\mathbb{Z}_2 \times \mathbb{Z}_2$, and one $\mathbb{Z}_2$ is this glide-symmetric $Z_2$ topological invariant, while the other $\mathbb{Z}_2$ is the Chern number in ***13*** and a half of the Chern number in ***15*** modulo 2 along the normal vector of the glide plane. On the other hand, in ***14***, the symmetry-based indicator is $\mathbb{Z}_2$, and we show that this $\mathbb{Z}_2$ is equal to the sum of the glide-symmetric $Z_2$ topological invariant and a half of the Chern number. It is interesting that the symmetry-based indicator gives a combination of two different topological invariants, while it does not uniquely determine the values of the individual topological invariants. Here we note that in ***14*** and ***15*** the Chern number on the glide plane is always an even integer.

Furthermore, we show that both the glide-symmetric $Z_2$ magnetic topological crystalline insulator in ***13*** and ***15*** is a higher-order topological insulator. In ***14***, the glide-symmetric $Z_2$ magnetic topological crystalline insulator with the Chern number equal to an integer multiple of four is a higher-order topological insulator. In ***14***, the Chern insulator with the Chern number equal to $4n + 2$ ($n$: integer) and the trivial glide-$Z_2$ invariant is also a higher-order topological insulator. Nevertheless, in the Chern insulators the surface is gapless, and one cannot call the latter case in ***14*** a higher-order topological insulator in a strict sense because the gapless hinge states are always hidden behind the gapless surface states due to the nonzero Chern number.

Our results will be useful for gaining insights for relating various topological phases. So far, various topological phases have been discovered by means of powerful methods, while their mutual relationships are not obvious in general. For example, we have shown that the glide-symmetric $Z_2$ topological crystalline insulator becomes the higher-order topological insulator in the presence of inversion symmetry, but such an equivalence is far from obvious from the known formula (4.3). Because the glide symmetry is one of the fundamental symmetries in crystals, and it is contained in many space groups, the present study will provoke studies on relating various topological invariants in some space groups and those in their supergroups.

## Appendix 1: Details of Calculations

### *1.1 Details of Rewriting the Berry Phase to Rotation/Screw Eigenvalues*

In Eqs. (4.28) and (4.42), we related some formulas containing the Berry phase to rotation or screw eigenvalues. This can be calculated by using sewing matrices we introduced in Sect. 4.1.2 following Ref. [9] without detailed calculations. As an example, here we show the details of Eq. (4.28). The integral of the Berry connection can be transformed to the determinant of the path-ordered Berry phase $\Gamma \to Y'(Y) \to \Gamma$,

$$\exp\left[-i\oint_{-\pi}^{\pi} dk_x \mathrm{tr}\mathcal{A}_x^-(k_x,0,0)\right] = \det\left[\mathcal{U}_{\Gamma Y'}^-\mathcal{U}_{Y\Gamma}^-\right], \tag{4.95}$$

where $\mathcal{U}_{\mathbf{k}_1\mathbf{k}_2}$ is the path-ordered Berry phase defined in Eq. (4.7), the superscript $(-)$ denotes the path-ordered Berry phase calculated from $\mathcal{A}^-(\mathbf{k})$, and the high-symmetry points $\Gamma$, $Y$, and $Y'$ are specified in Fig. 4.1. By using an orbital space operator in Eq. (4.8) and using the relation $\tilde{C}_2\tilde{U}_{\mathbf{k}_1\mathbf{k}_2}\tilde{C}_2^- = \tilde{U}_{C_2\mathbf{k}_1C_2\mathbf{k}_2}$, we have

$$\begin{aligned}
\det\left[\langle u_i(\Gamma)|\tilde{U}_\lambda|u_i(\Gamma)\rangle\right] &= \det\left[\langle u_i(\Gamma)|\tilde{U}_{\Gamma Y'}^-\tilde{U}_{Y\Gamma}^-|u_i(\Gamma)\rangle\right] \\
&= \det\left[\langle u_i(\Gamma)|\tilde{U}_{\Gamma Y'}^-\tilde{C}_2^{-1}\left(\tilde{C}_2\tilde{U}_{Y\Gamma}^-\tilde{C}_2^{-1}\right)\tilde{C}_2|u_i(\Gamma)\rangle\right] \\
&= \det\left[\langle u_i(\Gamma)|\tilde{U}_{\Gamma Y'}^-\tilde{C}_2^{-1}\tilde{U}_{Y'\Gamma}^-\tilde{C}_2|u_i(\Gamma)\rangle\right] \\
&= \det\Bigg[\sum_{m,n,l\in\mathrm{occ}} \langle u_i(\Gamma)|\tilde{U}_{\Gamma Y'}^-|u_m(Y')\rangle\langle u_m(Y')|\tilde{C}_2^{-1}|u_n(Y')\rangle \\
&\qquad \times\langle u_n(Y')|\tilde{U}_{Y'\Gamma}^-|u_l(\Gamma)\rangle\langle u_l(\Gamma)|\tilde{C}_2|u_i(\Gamma)\rangle\Bigg] \\
&= \det\left[\mathcal{U}_{\Gamma Y}^-\left(w_{C_2}^-(Y')\right)^{-1}\mathcal{U}_{Y'\Gamma}^- w_{C_2}^-(\Gamma)\right] \\
&= \det\left[\left(w_{C_2}^-(Y')\right)^{-1} w_{C_2}^-(\Gamma)\right]\det\left[\mathcal{U}_{\Gamma Y'}^-\mathcal{U}_{Y'\Gamma}^-\right] \\
&= \det\left[\left(w_{C_2}^-(Y)\right)^{-1} w_{C_2}^-(\Gamma)\right] = \prod_{i\in\mathrm{occ}}\frac{\zeta_i^-(\Gamma)}{\zeta_i^-(Y)},
\end{aligned}$$

where $w_{C_2}^-(\Gamma)$ is the sewing matrix for the $C_2$ operator, restricted within the $g_-$ subspace at the high-symmetry point $P$, and $\zeta_i^-(P)$ is the $C_2$ eigenvalue for the $g_-$ sector at the high-symmetry point $P$. We have used the periodicity along $k_x$. The identity operator we inserted $\sum_{m\in\mathrm{occ}}|u_m(\mathbf{k})\rangle\langle u_m(\mathbf{k})|$ is allowed if and only if $C_2$ rotation is a symmetry of the system and the system is fully gapped.

## 1.2 The Berry Connections and the Berry Curvatures on the $k_y = \pi$ Plane in *9*

Here we give details of the Berry connections and the Berry curvatures in ***9*** from the eigenstates in Eq. (4.56). The Berry connections $\tilde{\mathcal{A}}_x^{\pm}(k_x,\pi,k_z)$ and $\tilde{\mathcal{A}}_z^{\pm}(k_x,\pi,k_z)$ are given by

$$\begin{aligned}\tilde{\mathcal{A}}_x^{\pm}(k_x,\pi,k_z) &= \langle \tilde{u}^{\pm}(k_x,\pi,k_z)|i\partial_{k_x}|\tilde{u}^{\pm}(k_x,\pi,k_z)\rangle \\ &= \frac{1}{2}\left(\mathcal{A}_x(k_x,\pi,k_z)+\mathcal{A}_x(k_x,-\pi,k_z)\right)+\frac{1}{2\pi}\left(\chi(\pi,\pi,k_z)+\frac{k_z}{2}\right)-\frac{1}{2}\frac{\partial\chi(k_x,\pi,k_z)}{\partial k_x}-\frac{1}{4} \\ &\quad \pm\left(\frac{1}{8}e^{i(\chi(k_x,\pi,k_z)+k_z/2)}\langle u(k_x,\pi,k_z)|u(k_x,-\pi,k_z)\rangle+\text{c.c.}\right)+\begin{cases}0 & (g_+\text{ sector})\\ \frac{1}{2} & (g_-\text{ sector})\end{cases},\end{aligned}$$

$$\begin{aligned}\tilde{\mathcal{A}}_z^{\pm}(k_x,\pi,k_z) &= \langle \tilde{u}^{\pm}(k_x,\pi,k_z)|i\partial_{k_z}|\tilde{u}^{\pm}(k_x,\pi,k_z)\rangle \\ &= \frac{1}{2}\left(\mathcal{A}_z(k_x,\pi,k_z)+\mathcal{A}_z(k_x,-\pi,k_z)+\frac{k_x}{\pi}\frac{\partial\chi(\pi,\pi,k_z)}{\partial k_z}-\frac{\partial\chi(k_x,\pi,k_z)}{\partial k_z}+\frac{k_x}{2\pi}-\frac{1}{2}\right),\end{aligned}$$

respectively. Noticing that

$$e^{i(\chi(k_x,\pi,k_z)+k_z/2)}\langle u(k_x,\pi,k_z)|u(k_x,-\pi,k_z)\rangle = e^{ik_z/2}\langle u(k_x,\pi,k_z)|G(k_x,\pi,k_z)|u(k_x,-\pi,k_z)\rangle,$$

the Berry curvature $\mathcal{F}_{zx}^{\pm}(k_x,\pi,k_z)$ is therefore expressed as

$$\begin{aligned}\tilde{\mathcal{F}}_{zx}^{\pm}(k_x,\pi,k_z) &= \partial_{k_z}\tilde{\mathcal{A}}_x^{\pm}(k_x,\pi,k_z)-\partial_{k_x}\tilde{\mathcal{A}}_z^{\pm}(k_x,\pi,k_z)+i\left[\tilde{\mathcal{A}}_z^{\pm}(k_x,\pi,k_z)\tilde{\mathcal{A}}_x^{\pm}(k_x,\pi,k_z)\right] \\ &= \frac{1}{2}\left(\mathcal{F}_{zx}(k_x,\pi,k_z)+\mathcal{F}_{zx}(k_x,-\pi,k_z)\right) \\ &\quad \pm\frac{1}{8}\frac{\partial}{\partial k_z}\left(e^{ik_z/2}\langle u(k_x,\pi,k_z)|G(k_x,\pi,k_z)|u(k_x,-\pi,k_z)\rangle+\text{c.c.}\right).\end{aligned}$$

From these relations, Eqs. (4.57) and (4.58) directly follow.

## References


1. Aroyo MI, Perez-Mato JM, Capillas C, Kroumova E, Ivantchev S, Madariaga G, Kirov A, Wondratschek H (2006) Zeitschrift für Kristallographie-Crystalline Materials 221(1):15
2. Hahn T (2002) International tables for crystallography, volume a: space group symmetry, 5th edn. International Tables for Crystallography. Kluwer Academic Publishers, Dordrecht, Boston, London
3. Fang C, Fu L (2015) Phys Rev B 91:161105(R). https://doi.org/10.1103/PhysRevB.91.161105
4. Shiozaki K, Sato M, Gomi K (2015) Phys Rev B 91:155120. https://doi.org/10.1103/PhysRevB.91.155120
5. Shiozaki K, Sato M, Gomi K (2018) arXiv preprint arXiv:1802.06694
6. Po HC, Vishwanath A, Watanabe H (2017) Nat Commun 8(1):50
7. Watanabe H, Po HC, Vishwanath A (2018) Sci Adv 4(8):eaat8685
8. Ono S, Watanabe H (2018) Phys Rev B 98:115150. https://doi.org/10.1103/PhysRevB.98.115150
9. Fang C, Gilbert MJ, Bernevig BA (2012) Phys Rev B 86:115112. https://doi.org/10.1103/PhysRevB.86.115112
10. Song Z, Zhang T, Fang Z, Fang C (2018) Nat Commun 9(1):3530
11. Murakami S (2007) New J Phys 9(9):356
12. Wan X, Turner AM, Vishwanath A, Savrasov SY (2011) Phys Rev B 83:205101. https://doi.org/10.1103/PhysRevB.83.205101
13. Hughes TL, Prodan E, Bernevig BA (2011) Phys Rev B 83:245132. https://doi.org/10.1103/PhysRevB.83.245132

# Chapter 5
# Topological Invariants and Tight-Binding Models from the Layer Constructions

We construct all invariants which characterize topology of layer constructions (LCs) for systems with a given space group, in a similar way as in Ref. [1] in the present chapter. This construction is solely based on real-space geometries of layers. In Ref. [1], a LC is introduced for time-reversal-invariant systems, and in this previous work, each layer is decorated with a two-dimensional (2D) topological insulator (TI) and a 2D mirror-symmetric topological crystalline insulator (TCI). In contrast, in the present chapter, because we do not assume time-reversal symmetry (TRS), we decorate each layer with a 2D Chern insulator. We then construct all invariants based on each space-group operation, based solely on geometric properties of the layers. We find that these sets of invariants constructed by the LCs are in complete agreement with the topological invariants defined on $k$-space geometry discussed in the previous chapter. Thereafter, we give tight-binding models for those topological phases which consist of elementary layer constructions (eLCs). By applying our new formulas in the previous chapter to these simple tight-binding models, we confirm that the results are completely consistent with our theory. Besides, we support this conclusion by numerical calculations from our simple models.

## 5.1 Topological Invariants from the Layer Construction for Class A

In this section, we consider a LC in class A for ***7***, ***9***, ***13***, ***14***, and ***15***. We decorate each layer with a 2D Chern insulator because TRS is not enforced. Then we construct all invariants based solely on geometric properties of the layers for each space-group operation in a given space group. Then we show that these set of invariants are totally consistent with the topological invariants based on $k$-space geometry discussed in the previous chapter. Based on these arguments, we show eLCs, which constitute a basis for the sets of topological invariants.



H. Kim, *Glide-Symmetric $Z_2$ Magnetic Topological Crystalline Insulators*,
Springer Theses, https://doi.org/10.1007/978-981-16-9077-8_5

### 5.1.1 Setup

A layer $(mnl; d)$ is a set of planes given by the Miller indices $(mnl)$ displaced from the origin by $d$ $(0 \leq d < 1)$. It is given by

$$(mnl; d) = \{\mathbf{r}|\mathbf{r}\cdot(m\mathbf{b}_1 + n\mathbf{b}_2 + l\mathbf{b}_3) = 2\pi(d+q),\ q \in \mathbb{Z}\}$$
$$= \left\{\mathbf{r}|\frac{mx}{a} + \frac{ny}{b} + \frac{lz}{c} = (d+q),\ q \in \mathbb{Z}\right\}, \tag{5.1}$$

where $\mathbf{b}_i$'s are reciprocal vectors corresponding to the conventional lattice vectors $\mathbf{a}_1 = (1, 0, 0)$, $\mathbf{a}_2 = (0, 1, 0)$, $\mathbf{a}_3 = (0, 0, 1)$ and the integer $q$ is introduced because of the translation symmetry. Note that the primitive lattice vectors are equal to the conventional lattice vectors in primitive lattice systems whereas they are not in non-primitive lattice systems.

We then formulate general LCs, in the similar way as in Ref. [1]. In a given space group $\mathcal{G}$, symmetry property of a layer $L$ is described by its little group, which is defined as a subgroup of $\mathcal{G}$ that leaves $L$ invariant:

$$\mathcal{S}(L) = \{s \in \mathcal{G}|sL = L\}. \tag{5.2}$$

Then, $\mathcal{S}$ must also be a space group containing the full translation subgroup since by definition the layers are invariant under translation by any lattice vector would translate the layer. We can get a finite number of cosets of $\mathcal{S}$ for the space group $\mathcal{G}$,

$$\mathcal{G} = g_0\mathcal{S} + g_1\mathcal{S} + \cdots. \tag{5.3}$$

Thus, we have a set of symmetric layers

$$\{g_0L, g_1L, \ldots\} \tag{5.4}$$

by applying all the coset representatives on $L$. An eLC is obtained by decorating the layers in $\{g_0L, g_1L, \ldots\}$ with a 2D Chern insulator with a Chern number $+1$.

### 5.1.2 Topological Invariants for Glide-Symmetric Systems with Inversion Symmetry

First of all, we consider how a general layer $(mnl; d)$ gives topological invariants in the system with glide symmetry only, i.e., ***7*** and ***9***. Then, we expand our argument to ***13***, ***14***, and ***15*** by adding inversion symmetry. Then, inversion symmetry and either twofold ($C_2$) rotational symmetry or screw symmetry are added as generators. We briefly explain how topological invariants in the respective space groups can be calculated by these symmetries. Here we construct topological invariants associated with respective symmetry operations. We focus on one of the symmetry operations in the given space group, and in constructing a topological invariant associated with the symmetry operation, we neglect all the other symmetry operations in the space group. We then construct possible LCs compatible with the space group, and see whether each LC can be trivialized by deforming it while keeping its invariance under the focused symmetry operation. For example, in considering a glide invariant $\delta_{\mathrm{g}}$, we construct LCs satisfying the glide operation only, and then deform the LCs while keeping the glide symmetry, to see whether the LCs can be trivialized. This consideration leads to the definition of the glide topological invariant $\delta_{\mathrm{g}}$.

#### 5.1.2.1 Chern Invariant

The Chern invariant along a reciprocal lattice plane $\mathcal{D}_i$ normal to one of the *primitive* vectors $\mathbf{a}_{i=1,2,3}$ is written as

$$\delta_{C_i} = \frac{1}{2\pi}\int_{\mathcal{D}_i} d^2\mathbf{k}\; \mathbf{n}_i \cdot \mathbf{\Omega}_{\mathbf{k}}, \tag{5.5}$$

where $\mathbf{n}_i \equiv \mathbf{a}_i/|\mathbf{a}_i|$ is a unit vector along $\mathbf{a}_i$, $\mathbf{\Omega}_{\mathbf{k}}$ is the summation of the Berry curvature for all of the occupied states, and the integral is taken over the 2D Brillouin zone (BZ) on the plane $\mathcal{D}_i$. For a layer $L = (mnl; d)$, it is equal to the number of intersections of $L = (mnl; d)$ with an axis parallel to the $\mathbf{a}_i$ within the unit cell (Fig. 5.1a). For instance, one can easily see its intersections with $\mathbf{a}_{1,2,3}$ to be given by

$$\frac{d+q}{m}, \quad \frac{d+q}{n}, \quad \frac{d+q}{l}, \tag{5.6}$$

respectively. Thus, the Chern invariants $\delta_{C_{i=1,2,3}}$ associated with three primitive lattice vectors $\mathbf{a}_{1,2,3}$ should be given by

$$\delta_{C_1}(E) = \sum_{L\in E} m_L, \tag{5.7}$$

$$\delta_{C_2}(E) = \sum_{L\in E} n_L, \tag{5.8}$$

$$\delta_{C_3}(E) = \sum_{L\in E} l_L, \tag{5.9}$$

where $E$ is a general LC. In general, it can be written as

$$\delta_{C_i}(E) = \frac{1}{2\pi}\sum_{L\in E} \mathbf{g}_L \cdot \mathbf{a}_i, \tag{5.10}$$

where $\mathbf{g}_L = m\mathbf{b}_1 + n\mathbf{b}_2 + l\mathbf{b}_3$ for the layer $L \in E$.

#### 5.1.2.2 Glide Invariant $\delta_\mathrm{g}$

In addition, we can find a glide invariant for glide-symmetric systems based on the LC in the similar way as in time-reversal-invariant systems treated in Ref. [1]. In glide-invariant systems, there are two glide planes in a unit cell, which are displaced by a half of the primitive translation vector. In constructing the glide invariant $\delta_\mathrm{g}$, we should first choose one of the glide planes, and we fix this choice throughout our theory.

In our case each layer is decorated with the 2D Chern insulator because TRS is not assumed, whereas in Ref. [1], each layer is decorated with the 2D TIs. Nonetheless, nontrivial LC configurations are similar, and are classified into two types. In the first type of the nontrivial LC configuration, the glide plane is included in one of the layers in the LC (Fig. 5.1d). It is characterized by the glide-occupation number (glide-ON) $N^{\mathrm{o}}_{m,\mathbf{t}_\parallel}(L)$ where $m$ specifies the glide plane. In the second type of the nontrivial LC configuration, there is a pair of layers which are connected by the glide operation (Fig. 5.1b), which can be continuously deformed into the previous case (see Fig. 5.1c). It is represented by a glide-stacking number (glide-SN) $N^{\mathrm{s}}_{m,\mathbf{t}_\parallel}(L)$. Similarly to the cases in Ref. [1], two pairs of such layers can be trivialized. Therefore, the resulting topological invariant is defined modulo 2, and one can write the glide invariant $\delta_\mathrm{g}$ as

$$\delta_\mathrm{g}(E) = \sum_{L\in E}(N^{\mathrm{o}}_{m,\mathbf{t}_\parallel}(L) + N^{\mathrm{s}}_{m,\mathbf{t}_\parallel}(L)) \quad (\mathrm{mod}\ 2), \tag{5.11}$$

$$N^{\mathrm{o}}_{m,\mathbf{t}_\parallel}(L) = \begin{cases} 1 \text{ if } m \in L \\ 0 \text{ otherwise} \end{cases}, \tag{5.12}$$

$$N^{\mathrm{s}}_{m,\mathbf{t}_\parallel}(L) = \frac{1}{2\pi}|\mathbf{t}_\parallel \cdot \mathbf{g}_L|. \tag{5.13}$$

We again emphasize that in Eq. (5.12), $m$ refers to the glide plane which we have chosen from the two different choices of the glide plane.

#### 5.1.2.3 Inversion Invariant $\delta_\mathrm{i}$

Systems with inversion symmetry have eight inversion centers within a unit cell. We first choose one of the inversion centers, and we fix the choice throughout our theory. If a layer in the LC includes the inversion center (Fig. 5.2a), this LC cannot be trivialized as long as inversion symmetry is preserved. Meanwhile, if the inversion center is occupied by two layers decorated with the 2D Chern insulator, one can trivialize this configuration up to the creation of the atomic insulator at the center of inversion. Thus the inversion invariant $\delta_\mathrm{i}$ can be written as a inversion-ON $N^{\mathrm{o}}_i(L)$ of $L$ modulo 2:

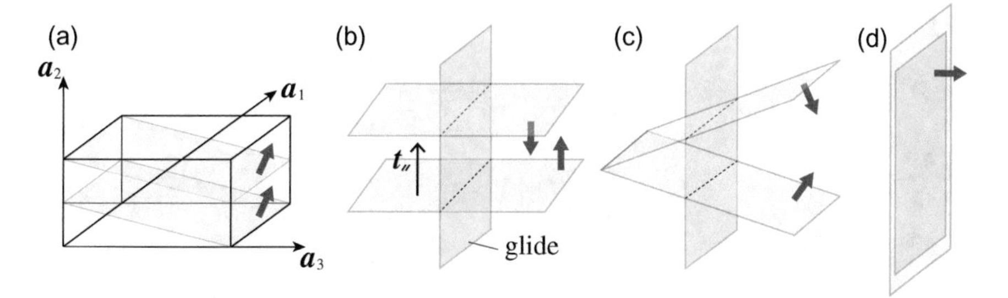


**Fig. 5.1** Chern invariants $\delta_{C_{i=1,2,3}}$ and glide invariant $\delta_{\mathrm{g}}$ in layer constructions. The yellow planes are the planes in the layer construction, and the blue planes are glide planes. The red arrows represent the directions of the reciprocal lattice vector $\mathbf{g}_L = m\mathbf{b}_1 + n\mathbf{b}_2 + l\mathbf{b}_3$, which specifies the orientation of the normal vector of the plane giving a positive Chern number $+1$. **a** Chern invariant $\delta_{C_i}$ $(i = 1, 2, 3)$ is determined as the number of intersections between the planes in the layer with an axis parallel to the $\mathbf{a}_i$ within the unit cell. The figure shows the case with $(021; 0)$, which intersects with $\mathbf{a}_1, \mathbf{a}_2$, and $\mathbf{a}_3$ by 0, 2, and 1 times per unit cell and yields $\delta_{C_1} = 0, \delta_{C_2} = 2$, and $\delta_{C_3} = 1$. **b–d** Glide invariant $\delta_{\mathrm{g}}$. **b** shows one of the nontrivial configurations of the layer. Two planes are perpendicular to the glide plane, and they are related by the glide operation, which means that they are separated by a glide vector $\mathbf{t}_{\parallel}$. Note that $2\mathbf{t}_{\parallel}$ is a lattice translation vector. **c** The configuration in **b** can be deformed into that in **d**. **d** shows another nontrivial configuration, where a plane in the layer matches with the glide plane

$$\delta_{\mathrm{i}}(E) = \sum_{L \in E} N_i^{\mathrm{o}}(L) \pmod 2, \tag{5.14}$$

$$N_i^{\mathrm{o}}(L) = \begin{cases} 1 \text{ if } i \in L \\ 0 \text{ otherwise} \end{cases}. \tag{5.15}$$

#### 5.1.2.4 $C_2$ Rotation Invariant $\delta_{\mathrm{r}}$

In a similar manner, the $C_2$ rotation invariant $\delta_{\mathrm{r}}$ is obtained. A layer of the 2D Chern insulator is $C_2$ rotation invariant only when the plane is perpendicular to the $C_2$ axis (Fig. 5.2b), and this layer can be trivialized, as we see from Fig. 5.2c. Consequently, the $C_2$ rotation invariant is always trivial:

$$\delta_{\mathrm{r}}(E) = 0. \tag{5.16}$$

It is a remarkable result, because the $C_2$ rotation invariant is always trivial in the system without TRS, while it is not in the presence of TRS [1]. In the presence of TRS, 2D TI layers are used for LCs, and the $C_2$ rotation invariant becomes nontrivial when a TI layer includes the $C_2$ axis. On the other hand, in our case where TRS is not enforced, layers of a 2D Chern insulator are used. In this case, the configuration where a layer of a 2D Chern insulator includes the $C_2$ axis is not $C_2$ invariant, unlike the case with TRS. Thus, there is no topologically nontrivial configuration for the $C_2$ rotation invariant.

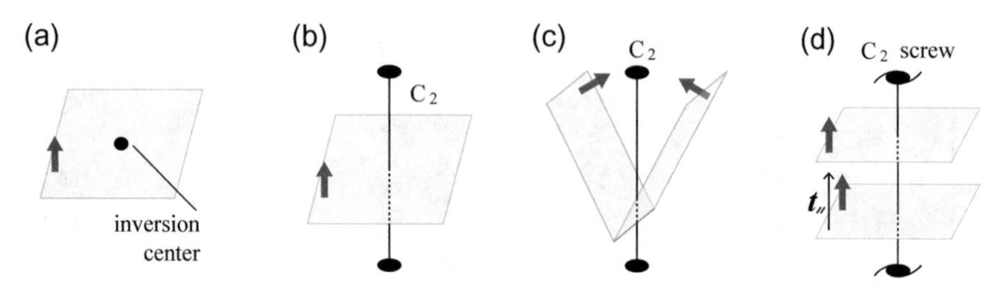


**Fig. 5.2** Inversion invariant $\delta_{\mathrm{i}}$, $C_2$ rotation invariant $\delta_{\mathrm{r}}$, and $C_2$ screw invariant $\delta_{\mathrm{s}}$ in layer constructions. **a** Inversion invariant $\delta_{\mathrm{i}}$. The plane includes the inversion center, which cannot be trivialized. **b**, **c** $C_2$ rotation invariant $\delta_{\mathrm{r}}$. In **b**, the plane is perpendicular to the $C_2$ rotation axis. This configuration is trivial, because as shown in **c** it can be trivialized while keeping the $C_2$ rotation symmetry. **d** $C_2$ screw invariant $\delta_{\mathrm{s}}$. The two planes in the layer are perpendicular to the $C_2$ screw axis, and they are related by the screw operation. $\mathbf{t}_{\parallel}$ is a screw vector, and is a half of a lattice translation vector. This configuration is nontrivial

#### 5.1.2.5 $C_2$ Screw Invariant $\delta_{\mathrm{s}}$

In systems with $C_2$ screw symmetry the topological invariant is defined by a screw-SN. The nontrivial LC configuration is the LC with a pair of layers combined by the screw operation. In such cases, one pair of layers combined by screw symmetry cannot be trivialized (Fig. 5.2d), while two pairs can be trivialized. Thus we get

$$\delta_{\mathrm{s}}(E) = \sum_{L\in E} N^{\mathrm{s}}_{s,\mathbf{t}_{\parallel}}(L) \pmod 2, \tag{5.17}$$

$$N^{\mathrm{s}}_{s,\mathbf{t}_{\parallel}}(L) = \frac{1}{2\pi}|\mathbf{t}_{\parallel}\cdot \mathbf{g}_L|, \tag{5.18}$$

where $\mathbf{t}_{\parallel}$ is a screw vector, and is equal to a half of a lattice translation vector.

### 5.1.3 *Invariants for Layer Constructions*

Based on the definitions of the invariants given above, one can calculate their values for LCs in *7*, *9*, *13*, *14*, and *15*. We note that in glide- and inversion-symmetric systems, we choose the glide planes and inversion centers to be the standard ones given in the previous chapter, which are the same as those in the Bilbao Crystallographic Server [2].

#### 5.1.3.1 Space Group *7*

In *7*, only the Chern invariants $\delta_{C_{i=1,2,3}}$ and the glide invariant $\delta_{\mathrm{g}}$ are defined. In the calculation we note that the glide operation $G_y = \{m_{010}|\mathbf{a}_3/2\}$ transforms the layer

$L = (mnl; d)$ to

$$\begin{aligned} G_y L &= \left\{ g_y \mathbf{r} | \frac{m}{a}x - \frac{n}{b}y + \frac{l}{c}\left(z + \frac{c}{2}\right) = d + q,\ q \in \mathbb{Z} \right\} \\ &= \left\{ g_y \mathbf{r} | \frac{(-m)}{a}x + \frac{n}{b}y + \frac{(-l)}{c}z = \frac{l}{2} - d - q,\ q \in \mathbb{Z} \right\} \\ &= \left( \bar{m}n\bar{l} \middle| \frac{l}{2} - d \right), \end{aligned} \tag{5.19}$$

where $g_y \mathbf{r}$ is transformed from $\mathbf{r}$ by $G_y$. Note that it is not written as $(m\bar{n}l|d - \frac{l}{2})$. Therefore, the Chern invariants for the LC $L(mnl; d)$ in *7* are given as

$$\delta_{C_2}(E) = \sum_{L \in E} N_C(L) \quad (\mathrm{mod}\ \mathbb{Z}), \tag{5.20}$$

$$N_C(L) = \begin{cases} n_L & (m_L = 0,\ l_L = 0,\ d_L = 0, 1/2) \\ 2n_L & (\text{otherwise}) \end{cases}, \tag{5.21}$$

$$\delta_{C_1}(E) = 0, \tag{5.22}$$

$$\delta_{C_3}(E) = 0, \tag{5.23}$$

where the classification is necessary for distinguishing the cases $L = G_y L$ and $L \neq G_y L$. Here $\delta_{C_1}(E) = 0$ and $\delta_{C_3}(E) = 0$ follow from glide symmetry.

The glide invariant is given by

$$\delta_g \equiv \begin{cases} 1 & (m_L = 0,\ l_L = 0,\ d_L = 0) \\ l_L & (\text{otherwise}) \end{cases} \tag{5.24}$$

modulo 2. We note that the glide plane is chosen to be $y = 0$, and not $y = 1/2$. That is why (5.24) does not contain $d_L = 1/2$ in the first case of $\delta_g = 1$.

#### 5.1.3.2 Space Group *13*

In *13*, we can define four kinds of invariants, the Chern invariants $\delta_{C_{i=1,2,3}}$, the glide invariant $\delta_g$, the inversion invariant $\delta_i$, and the $C_2$ rotation invariant $\delta_r$, only from geometric configuration of the layer constructed from *7* by adding inversion symmetry. The Chern invariant $\delta_{C_i}$ takes an integer value, while the other invariants $\delta_g$ and $\delta_i$ are defined modulo 2, and $\delta_r$ turns out to be always trivial. The inversion transforms the layer $(mnl; d)$ into $(mnl; -d)$. This additional inversion symmetry doubles the number of layers, if the additional layers are not identical with the original ones, Then for every LC, one can evaluate these invariants, from which we establish relations between these invariants. For the layer $L(mnl; d)$, the values of the Chern invariants are obtained as follows:

$$\delta_{C_1} = 0, \tag{5.25}$$
$$\delta_{C_3} = 0, \tag{5.26}$$
$$\delta_{C_2} = \begin{cases} n_L \ \ (m_L = 0,\ l_L = 0,\ d_L = 0, 1/2) \\ 2n_L \ (m_L = 0,\ l_L = 0,\ d_L \neq 0, 1/2) \\ 2n_L \ ((m_L, l_L) \neq (0,0),\ d_L = 0, 1/2) \\ 4n_L \ ((m_L, l_L) \neq (0,0),\ d_L \neq 0, 1/2) \end{cases}. \tag{5.27}$$

The glide invariant and the inversion invariant are calculated, and they turn out to be exactly equal;

$$\delta_{\mathrm{g}} = \delta_{\mathrm{i}} \equiv \begin{cases} 1 \ \ (m_L = 0,\ l_L = 0,\ d_L = 0) \\ l_L \ ((m_L, l_L) \neq (0,0),\ d_L = 0, 1/2) \\ 0 \ \ (\text{otherwise}) \end{cases} \tag{5.28}$$

modulo 2. The $C_2$ rotation invariant is always trivial:

$$\delta_{\mathrm{r}} \equiv 0 \pmod 2. \tag{5.29}$$

Now we can compare these invariants with the glide-$Z_2$ invariant $\tilde{\nu}$ and the Chern number $C_y$. One can calculate these two topological invariants for general layers and compare them with the invariants $\delta_{C_i}$, $\delta_{\mathrm{g}}$, $\delta_{\mathrm{i}}$, and $\delta_{\mathrm{r}}$ from the LC. Then we get

$$C_y = \delta_{C_2}, \tag{5.30}$$
$$\tilde{\nu} \equiv \delta_{\mathrm{i}} \equiv \delta_{\mathrm{g}} \pmod 2. \tag{5.31}$$

Thus the construction of the topological invariants based on the LC completely agrees with the known topological invariants, the glide-$Z_2$ invariant $\tilde{\nu}$ and the Chern number $C_y$. From these results we can demonstrate generators of LCs for nontrivial values of topological invariants, i.e., minimal layer configurations for nontrivial combinations of the topological invariants. They are listed in Table 5.1.

#### 5.1.3.3 Space Group *14*

One can define topological invariants based on the LC, as has been done for ***13***. In ***14***, we can define four kinds of invariants, the Chern invariants $\delta_{C_{i=1,2,3}}$, the glide invariant $\delta_{\mathrm{g}}$, the inversion invariant $\delta_{\mathrm{i}}$, and the $C_2$ screw invariant $\delta_{\mathrm{s}}$, from geometric configurations of the layers. The Chern invariant $\delta_{C_i}$ takes an integer value, while the other invariants $\delta_{\mathrm{g}}$, $\delta_{\mathrm{i}}$, and $\delta_{\mathrm{s}}$ are defined modulo 2. Then for every LC, one can evaluate these invariants, from which we establish relations between these invariants. At first, for the layer $L(mnl; d)$, the values of the Chern invariants are obtained as follows:

$$\delta_{C_1} = 0, \tag{5.32}$$

$$\delta_{C_3} = 0, \tag{5.33}$$

$$\delta_{C_2} = \begin{cases} n_L & (m_L = 0,\ n_L = \text{even},\ l_L = 0,\ d_L = 0, 1/2) \\ 2n_L & (m_L = 0,\ n_L = \text{odd},\ l_L = 0,\ d_L = 1/4, 3/4) \\ 2n_L & ((m_L, l_L) \neq (0,0),\ n_L = \text{odd},\ d = 0, 1/2) \\ 4n_L & (\text{otherwise}) \end{cases}. \tag{5.34}$$

Second, the glide invariant is given by

$$\delta_{\text{g}} \equiv \begin{cases} 1 \ (m_L = 0,\ l_L = 0,\ d_L \equiv \pm\frac{n_L}{4} \pmod 1) \\ 1 \ (l_L = \text{odd},\ d_L = 0, 1/2) \\ 0 \ (\text{otherwise}) \end{cases} \tag{5.35}$$

modulo 2. Next, the inversion invariant is given by

$$\delta_{\text{i}} \equiv \begin{cases} 1 \ (m_L = 0,\ n_L = \text{even},\ l_L = 0,\ d_L = 0) \\ 1 \ (n_L + l_L = \text{odd},\ d_L = 0, 1/2) \\ 0 \ (\text{otherwise}) \end{cases} \tag{5.36}$$

modulo 2. Finally, the $C_2$ screw invariant is given by

$$\delta_{\text{s}} \equiv \begin{cases} 1 \ (n_L = \text{odd},\ d_L = 0, 1/2) \\ 1 \ (m_L = 0,\ n_L = \text{odd},\ l_L = 0,\ d_L = 0, 1/4, 1/2, 3/4) \\ 1 \ (m_L = 0,\ n_L = 2 \times \text{odd},\ l_L = 0,\ l_L = 0, 1/2) \\ 0 \ (\text{otherwise}) \end{cases} \tag{5.37}$$

modulo 2. Therefore, they are related by the following equations:

$$\delta_{C_2} \in 2\mathbb{Z}, \tag{5.38}$$

$$\delta_{\text{i}} \equiv \delta_{\text{g}} + \frac{1}{2}\delta_{C_2} \pmod 2, \tag{5.39}$$

$$\delta_{\text{s}} \equiv \frac{1}{2}\delta_{C_2} \pmod 2. \tag{5.40}$$

Let us compare these invariants with the glide-$Z_2$ invariant $\tilde{\nu}$ and the Chern number $C_y$, and we get

$$\delta_{C_2} = C_y \in 2\mathbb{Z}, \tag{5.41}$$

$$\delta_{\text{g}} \equiv \tilde{\nu} \pmod 2, \tag{5.42}$$

$$\delta_{\text{i}} \equiv \tilde{\nu} + \frac{1}{2}C_y \pmod 2, \tag{5.43}$$

$$\delta_{\text{s}} \equiv \frac{1}{2}C_y \pmod 2. \tag{5.44}$$

Thus the topological invariants constructed from the LC completely agree with the known topological invariants, the glide-$Z_2$ invariant $\tilde{\nu}$ and the Chern number $C_y$. From these results we can demonstrate generators of LCs for nontrivial values of topological invariants, i.e., minimal layer configurations for nontrivial combinations of the topological invariants. They are listed in Table 5.1.

#### 5.1.3.4 Space Group *9*

In *9*, we can find the Chern invariants $\delta_{C_{i=1,2,3}}$, a glide invariant $\delta_{g_y}$ defined by the glide operation $G_y = \{m_{010}|\mathbf{a}_3/2\}$, and the other glide invariant $\delta_{g_n}$ defined by the glide operation $G_n = \{m_{010}|(\mathbf{a}_1 + \mathbf{a}_2 + \mathbf{a}_3)/2\}$.

The glide operation $G_n$ arises from a different nature of the symmetry operations from $G_y$ and $G_n(= TG_y)$ is known as *additional symmetry elements* [3], where $T$ is a symmetry translation. Here we explain the reason why the glide invariant $\delta_{g_n}$ arises following the arguments in Refs. [1, 3]. First, we recall the group theory to generate all the symmetry elements in a given space group. A general space group operation $\{p|\mathbf{t}\}$ consists of a point group operation $p$ and a translation $\mathbf{t}$: $\{p|\mathbf{t}\} : \mathbf{r} \to p\mathbf{r} + \mathbf{t}$. Let us consider a space group operation $\{p|\mathbf{t} + \mathbf{v}\}$. When we consider the translation $\mathbf{t}_{\parallel}$ parallel to a given symmetry axis or plane, under this operation we get

$$p\mathbf{r} + \mathbf{t} + \mathbf{v} = p(\mathbf{r} - \mathbf{x}) + \mathbf{x} + \mathbf{t}_{\parallel}, \tag{5.45}$$

where $p\mathbf{t}_{\parallel} = \mathbf{t}_{\parallel}$ and $\mathbf{x}$ is a center operated a point group operation. Note that the normal component of $\mathbf{t}$ is responsible for its location. If $\mathbf{v}$ has a nonzero normal component, the operation $\{p|\mathbf{t} + \mathbf{v}\}$ is in a different type from the original symmetry $\{p|\mathbf{t}\}$. It generates additional symmetry elements. These stem from nonprimitive unit cells and a fractional translation in screw symmetry or glide symmetry, and allow additional screw or glide symmetry which arises from rotation and screw axes or mirror and glide planes, respectively.

As we mentioned in Sect. 4.4, the unit cell of the monoclinic primitive lattice (*7*) corresponding the conventional unit cell is achieved by doubling that of the monoclinic base-centered lattice (*9*), and there exist additional symmetry elements derived from nonprimitive unit cells. Therefore we have to consider all the symmetry operations including additional symmetry elements to exhaust invariants from the LC defined in the conventional unit cell [1].

Let us investigate the invariants in *9*. The glide operations $G_y$ and $G_n$ transform the layer $L = (mnl; d)$ to

$$G_y L = \left(\bar{m}n\bar{l}\middle|\frac{l}{2} - d\right), \tag{5.46}$$

$$G_n L = \left(\bar{m}n\bar{l}\middle|\frac{1}{2}(m - n + l) - d\right), \tag{5.47}$$

and

$$G_n G_y^{-1} L = T_1 L = \left( mnl; d - \frac{1}{2}(m+n) \right), \tag{5.48}$$

where $T_1$ is the translation operation along the *primitive* lattice vector $\mathbf{a}_1 = (\frac{1}{2}\frac{1}{2}0)$ for ***9***. Note that $G_y L$ is not written as $(m\bar{n}l; d - \frac{l}{2})$ because the 2D Chern insulator has its own orientation corresponding to the direction of the chiral edge states. Thus, there are four types of layers in ***9***. Therefore, the Chern invariants for the LC $L(mnl; d)$ in ***9*** are given as

$$\delta_{C_1} = \delta_{C_2} = \sum_{L \in E} N_C(L) \quad (\text{mod } \mathbb{Z}), \tag{5.49}$$

$$N_C(L) = \begin{cases} n_L/2 \; (m_L = 0, n_L = \text{even}, l_L = 0, d_L = 0, 1/2) \\ n_L \quad (m_L = 0, n_L = \text{odd}, l_L = 0, d_L = 0, 1/2 \text{ or} \\ \qquad m_L = 0, l_L = 0, d_L = 1/4, 3/4 \text{ or } m_L + n_L = \text{even}) \\ 2n_L \; (\text{otherwise}) \end{cases}, \tag{5.50}$$

$$\delta_{C_3} = 0. \tag{5.51}$$

We emphasize that the Chern invariants $\delta_{C_{i=1,2,3}}$ are along the *primitive* lattice vectors for ***9***.[1]

The glide invariant $\delta_{g_y}$ is given by

$$\delta_{g_y} \equiv \begin{cases} 1 \; (m_L = 0, l_L = 0, d_L = 0) \\ l_L \; (m_L + n_L = \text{even}) \\ 0 \; (\text{otherwise}) \end{cases} \tag{5.52}$$

modulo 2. The glide plane is chosen to be $y = 0$, and not $y = 1/2$. This is why (5.52) does not contain $d_L = 1/2$ in the first case of $\delta_{g_y} = 1$. Similarly, the other glide invariant $\delta_{g_n}$ is given by

$$\delta_{g_n} \equiv \begin{cases} 1 \qquad (m_L = 0, n_L \in 4\mathbb{Z}, l_L = 0, d_L = 0 \text{ or} \\ \qquad\; m_L = 0, n_L = \text{odd}, l_L = 0, d_L = 1/4 \text{ or} \\ \qquad\; m_L = 0, n_L = 2 \times \text{odd}, l_L = 0, d_L = 1/2) \\ m_L + l_L \; (m_L + n_L = \text{even}) \\ 0 \qquad (\text{otherwise}) \end{cases} \tag{5.53}$$

modulo 2. Here we choose the glide plane to be $y = 1/4$, and not $y = 3/4$.

[1] Refer to Table 4.4.

#### 5.1.3.5 Space Group *15*

In ***15***, we can define six kinds of invariants, the Chern invariants $\delta_{C_{i=1,2,3}}$, the two different types of glide invariant $\delta_{g_y}$, $\delta_{g_n}$, the inversion invariant $\delta_{\mathrm{i}}$, the $C_2$ rotation invariant $\delta_{\mathrm{r}}$, and the $C_2$ screw invariant $\delta_{\mathrm{s}}$, only from geometric configuration of the layer constructed from ***9*** by adding inversion symmetry. The $C_2$ screw invariant is additionally defined in ***15*** due to the nonprimitive nature similar to ***9***. The Chern invariant $\delta_{C_i}$ takes an integer value, while the other invariants $\delta_{g_y}$, $\delta_{g_n}$, $\delta_{\mathrm{i}}$, and $\delta_{\mathrm{s}}$ are defined modulo 2, and $\delta_{\mathrm{r}}$ turns out to be always zero. The inversion operator transforms the layer $(mnl; d)$ into $(mnl; -d)$, and this leads to double the number of layers to be eight, if the additional layers are not identical with the original ones. In this system, the Chern invariants are obtained as follows:

$$\delta_{C_1} = \delta_{C_2} = \begin{cases} n_L/2 & (m_L = 0, n_L = \text{even}, l_L = 0, d_L = 0, 1/2) \\ n_L & (m_L = 0, n_L = \text{odd}, l_L = 0, d_L = 0, 1/4, 1/2, 3/4 \text{ or} \\ & \;\; m_L = 0, n_L = \text{even}, l_L = 0 \text{ or } m_L + n_L = \text{even}, d_L = 0, 1/2) \\ 2n_L & (m_L + n_L = \text{even or } d_L = 0, 1/2 \text{ or } m_L = 0, l_L = 0) \\ 4n_L & (\text{otherwise}) \end{cases}, \tag{5.54}$$

$$\delta_{C_3} = 0. \tag{5.55}$$

The glide invariant $\delta_{g_y}$ and the inversion invariant $\delta_{\mathrm{i}}$ are calculated, and they turn out to be completely equal;

$$\delta_{g_y} = \delta_{\mathrm{i}} \equiv \begin{cases} 1 & (m_L = 0, l_L = 0, d_L = 0) \\ l_L & (m_L + n_L = \text{even}, d_L = 0, 1/2) \\ 0 & \text{otherwise} \end{cases} \tag{5.56}$$

modulo 2. The other glide invariant $\delta_{g_n}$ is given by

$$\delta_{g_n} \equiv \begin{cases} 1 & (m_L = 0, n_L \in 4\mathbb{Z}, l_L = 0, d_L = 0 \text{ or} \\ & \;\; m_L = 0, n_L = \text{odd}, l_L = 0, d_L = 1/4 \text{ or} \\ & \;\; m_L = 0, n_L = 2 \times \text{odd}, l_L = 0, d_L = 1/2) \\ m_L + l_L & (m_L + n_L = \text{even}, d_L = 0, 1/2) \\ 0 & (\text{otherwise}) \end{cases} \tag{5.57}$$

modulo 2. The $C_2$ screw invariant $\delta_{\mathrm{s}}$ is given by

$$\delta_{\mathrm{s}} \equiv \begin{cases} n_L/2 & (m_L = 0, n_L = \text{even}, l_L = 0, d_L = 0, 1/2) \\ n_L & (m_L = 0, n_L = \text{odd}, l_L = 0, d = 0, 1/4, 1/2, 3/4 \text{ or} \\ & \;\; m_L + n_L = \text{even}, d_L = 0, 1/2) \end{cases} \tag{5.58}$$

modulo 2. They are related by the following equations:

$$\delta_{\mathrm{i}} \equiv \delta_{\mathrm{g}_y} \pmod 2, \tag{5.59}$$

$$\delta_{\mathrm{s}} \equiv \frac{1}{2}\delta_{C_y} \pmod 2, \tag{5.60}$$

$$\delta_{\mathrm{i}} + \delta_{\mathrm{s}} \equiv \delta_{\mathrm{g}_n} \pmod 2, \tag{5.61}$$

in which we define $\delta_{C_y} = \delta_{C_1} + \delta_{C_2} = \delta_{C_1}$.

Let us verify the relations between these invariants and the glide-$Z_2$ invariant $\tilde{\nu}$ and the Chern number $C_y$. We can calculate $\tilde{\nu}$ and $C_y$ for general layers and compare them with the invariants from the LC. We then get

$$\delta_{C_y} = C_y \in 2\mathbb{Z}, \tag{5.62}$$

$$\delta_{\mathrm{i}} \equiv \delta_{\mathrm{g}_y} \equiv \tilde{\nu} \pmod 2, \tag{5.63}$$

$$\delta_{\mathrm{s}} \equiv \frac{1}{2}C_y \pmod 2, \tag{5.64}$$

$$\delta_{\mathrm{g}_n} \equiv \tilde{\nu} + \frac{1}{2}C_y \pmod 2. \tag{5.65}$$

The topological invariants calculated from the LC hence totally agree with the known topological invariants, the glide-$Z_2$ invariant $\tilde{\nu}$ and the Chern number $C_y$. We will show minimal layer configurations for nontrivial combinations of the topological invariants from these results shown in Table 5.1.

### 5.1.4 Elementary Layer Constructions

Following the argument in Ref. [1], we can find eLCs in ***7***, ***9***, ***13***, ***14***, and ***15*** summarized in Table 5.1. All space groups listed here have two independent topological invariants as we have seen in this thesis. Therefore, each space group has two eLCs.

### 5.1.5 Convention Dependence of Topological Invariants

As remarked in Ref. [1], topological invariants in the previous subsection depend on the conventions. For example, there are eight inversion centers in a unit cell in the presence of inversion symmetry, and we should choose one inversion center among them in defining an inversion invariant $\delta_{\mathrm{i}}$. When we choose the inversion center at the origin, $(mnl; 0)$ passing the origin has $\delta_{\mathrm{i}} = 1$ whereas $(mnl; \frac{1}{2})$ has $\delta_{\mathrm{i}} = 0$ because it does not pass the inversion center. On the other hand, when we choose the inversion center at $(0, 0, \frac{1}{2})$, $(mnl; 0)$ has $\delta_{\mathrm{i}} = 0$ whereas $(mnl; \frac{1}{2})$ has $\delta_{\mathrm{i}} = 1$. Thus, the inversion invariant $\delta_{\mathrm{i}}$ depends on the convention on the choice of the inversion center among the eight inversion centers in the unit cell. Meanwhile, when we consider an LC with two eLC passing all eight inversion centers in a unit

**Table 5.1** Summary of elementary layer constructions, symmetry-based indicators, and topological invariants for ***7***, ***9***, ***13***, ***14***, and ***15***

| SG | eLC | $\mathbb{Z}_{2,2,2,4}$ | Invariants | | | | | |
|---|---|---|---|---|---|---|---|---|
| | $(mnl; d)$ | | $\delta_{C_i}$ | $g^{010}_{00\frac{1}{2}}$ | | | | |
| ***7*** $(Pc)$ | $001; d_0$ | N/A | 000 | 1 | | | | |
| | $010; 0$ | | 010 | 0 | | | | |
| | $(mnl; d)$ | | $\delta_{C_i}$ | $g^{010}_{00\frac{1}{2}}$ | $g^{010}_{\frac{1}{2}0\frac{1}{2}}$ | | | |
| ***9*** $(Cc)$ | $001; d_0$ | N/A | 000 | 1 | 1 | | | |
| | $\bar{1}10; d_0$ | | 110 | 0 | 1 | | | |
| | $(mnl; d)$ | | $\delta_{C_i}$ | $g^{010}_{00\frac{1}{2}}$ | $i$ | $2^{010}$ | | |
| ***13*** $(P2/c)$ | $001; 0$ | 0002 | 000 | 1 | 1 | 0 | | |
| | $010; \frac{1}{2}$ | 0100 | 010 | 0 | 0 | 0 | | |
| | $(mnl; d)$ | | $\delta_{C_i}$ | $g^{010}_{00\frac{1}{2}}$ | $i$ | $2^{010}_1$ | | |
| ***14*** $(P2_1/c)$ | $001; 0$ | 0002 | 000 | 1 | 1 | 0 | | |
| | $020; 0$ | 0002 | 020 | 0 | 1 | 1 | | |
| | $(mnl; d)$ | | $\delta_{C_i}$ | $g^{010}_{00\frac{1}{2}}$ | $g^{010}_{\frac{1}{2}0\frac{1}{2}}$ | $i$ | $2^{010}$ | $2^{010}_1$ |
| ***15*** $(C2/c)$ | $001; 0$ | 0002 | 000 | 1 | 1 | 1 | 0 | 0 |
| | $\bar{1}10; 0$ | 0100 | 110 | 0 | 1 | 0 | 0 | 1 |

cell, the inversion invariant is independent of the choice of the inversion center, and the system is considered to be topologically nontrivial. Similarly, the glide-ON is also dependent on the convention. There are two glide planes within the unit cell, displaced from each other by half the primitive translation vector, and the glide invariant $\delta_{\mathrm{g}}$ depends on the choice. When the LC occupies both of these glide planes odd times each, the glide invariant $\delta_{\mathrm{g}}$ is nontrivial, irrespective of the choice of the glide plane, as is similar to the cases with TRS in Ref. [1].

In our cases in ***13*** as an example, the configuration in Fig. 5.3a has a nontrivial inversion invariant $\delta_{\mathrm{i}}$, irrespective of the convention of the inversion center since the layers pass all the eight inversion centers in the unit cell. Meanwhile, in the configuration in Fig. 5.3b, the inversion invariant $\delta_{\mathrm{i}}$ depends on the convention. In ***14***, both of the two configurations in Fig. 5.9a, b have a convention-independent nontrivial inversion invariant $\delta_{\mathrm{i}}$.

## 5.2 Tight-Binding Models Constructed from the Layer Construction

In the present section, we construct tight-binding models representing eLCs in Table 5.1. In order to verify our theory in the previous chapter, we then calculate the

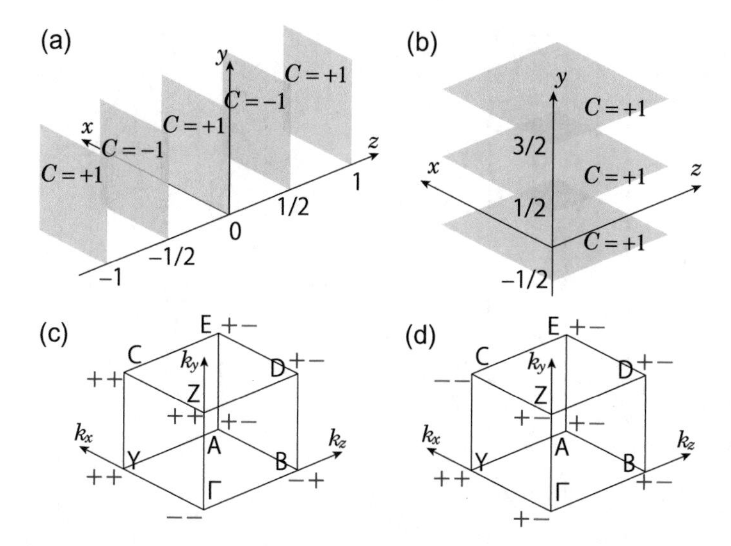


**Fig. 5.3** Layer constructions for **a** (001; 0), showing a glide-$Z_2$ topological phase ($\tilde{\nu} = 1, C_y = 0$) and **b** (010; $\frac{1}{2}$), showing a Chern insulator ($\tilde{\nu} = 0, C_y = 1$) with ***13***. The parities at high-symmetry points for the models **a** and **b** are shown in **c** and **d**, respectively

topological invariants by exploiting our new formulas. The results support our new formulas together with the argument of the LC in the previous section.

### 5.2.1 Tight-Binding Models Constructed from the Layer Construction for 13

First, we study the two minimal LCs (001; 0) and (010; $\frac{1}{2}$) in Table 5.1 for ***13***. These two LCs correspond to the glide-symmetric $Z_2$ magnetic TCI and the Chern insulator phases in ***13***, and in this subsection we construct simple tight-binding models for them in order to examine our scenario. We show the outline of the calculations in this section, and the details are presented in Appendix 1.1.

#### 5.2.1.1 Glide-Symmetric $Z_2$ Magnetic TCI Phase: (001; 0)

We begin with the eLC (001; 0) showing the glide-symmetric $Z_2$ magnetic TCI phase (Fig. 5.3a). In the absence of inversion symmetry, i.e., for ***7***, a minimal configuration for a glide-symmetric $Z_2$ magnetic TCI can be realized as a LC (001; $d_0$), representing 2D Chern insulators with $C_z = +1$ along the $xy$ plane placed at $z = z_0 + n$ where $n$ is an integer. Then, the glide symmetry requires presence of other layers of a 2D Chern insulator with $C_z = -1$ at $z = z_0 + n + 1/2$, because the glide

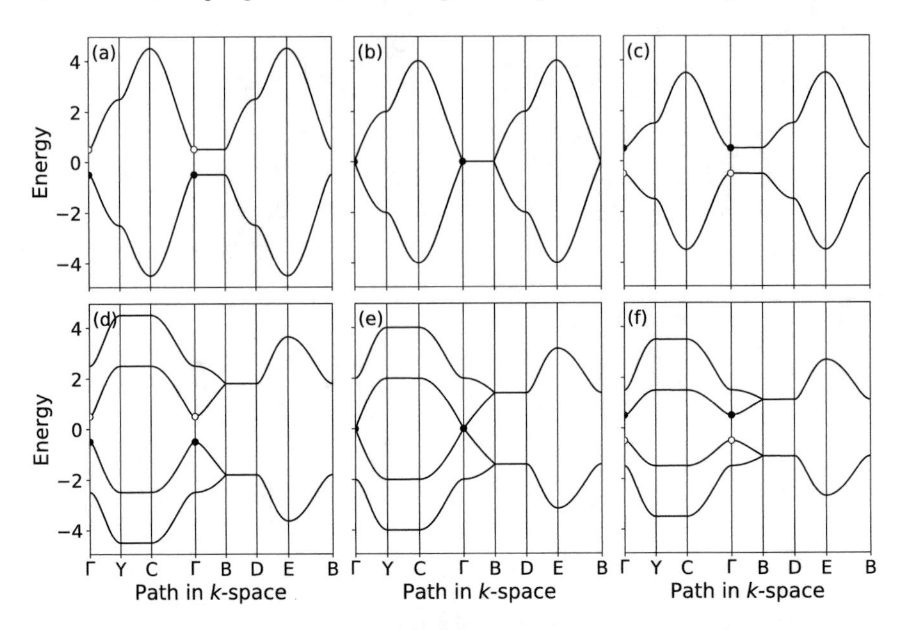


**Fig. 5.4** Bulk band structures for the tight-binding models for **a–c** the glide-symmetric $Z_2$ magnetic topological crystalline insulator phase of Eq. (5.66) and **d–e** the Chern insulator phase of Eq. (5.78) constructed from the layer construction in ***13*** with **a, d** $m = -2.5$; **b, e** $m = -2$; **c, f** $m = -1.5$. Filled (open) circles mark states of common character; band inversion at $\Gamma$ is evident in **c** and **f**

operation changes the sign of the Chern number along the $xy$ plane. This model gives a nontrivial value of the $Z_2$ topological invariant $\tilde{\nu} = 1$ for glide-symmetric systems by a direct calculation.

We now consider the case when inversion symmetry is added, and the space group becomes ***13***. Inversion symmetry fixes the value of $z_0$ to be $z_0 = 0$, and the configuration allowed by symmetry is the one with the 2D Chern insulators on $z = n$ with $C_z = +1$ and on $z = n + 1/2$ with $C_z = -1$, where $n$ is an integer (Fig. 5.3a). This is the LC with (001; 0). Since a representative lattice model exhibiting a single layer of the 2D Chern insulator is given by Eq. (2.118), we can decorate the 2D Chern insulators on $z = n$ as Eq. (2.118) and make copies by glide operation corresponding to the 2D Chern insulators on $z = n + 1/2$. Therefore, a representative Hamiltonian for the glide-symmetric $Z_2$ magnetic TCI phase with additional inversion symmetry is obtained as

$$H_{\tilde{\nu}}^{\mathrm{LC}}(\mathbf{k}) = (m + \cos k_x + \cos k_y)\sigma_z + \sin k_x \sigma_x + \sin k_y \sigma_y \tau_z, \tag{5.66}$$

where $\boldsymbol{\sigma}$ and $\boldsymbol{\tau}$ are Pauli matrices denoting the orbital and lattice degrees of freedom.[2] We here set the Fermi energy to be $E_F = 0$. Hence, the corresponding $k$-dependent glide operator and $k$-dependent inversion operator are

[2] Details are given in Sect. 5.3.

**Table 5.2** Irreducible representations for the two tight-binding models for the topological phases in ***13*** from the layer construction

| | $\Gamma$ | $Y$ | $Z$ | $C$ |
|---|---|---|---|---|
| $H_{\tilde{\nu}}^{\mathrm{LC}}(m=-2.5)$ | $\Gamma_1^+ + \Gamma_2^+$ | $Y_1^+ + Y_2^+$ | $Z_1^+ + Z_2^+$ | $C_1^+ + C_2^+$ |
| $H_{\tilde{\nu}}^{\mathrm{LC}}(m=-1.5)$ | $\Gamma_1^- + \Gamma_2^-$ | $Y_1^+ + Y_2^+$ | $Z_1^+ + Z_2^+$ | $C_1^+ + C_2^+$ |
| $H_{C}^{\mathrm{LC}}(m=-2.5)$ | $\Gamma_1^+ + \Gamma_2^+$ | $Y_1^+ + Y_2^+$ | $Z_1^- + Z_2^-$ | $C_1^- + C_2^-$ |
| $H_{C}^{\mathrm{LC}}(m=-1.5)$ | $\Gamma_2^+ + \Gamma_2^-$ | $Y_1^+ + Y_2^+$ | $Z_2^+ + Z_2^-$ | $C_1^- + C_2^-$ |

$$G_y(k_z) = e^{-ik_z/2}\left(\cos\frac{k_z}{2}\tau_x + \sin\frac{k_z}{2}\tau_y\right), \tag{5.67}$$

$$I(k_z) = \sigma_z \begin{pmatrix} 1 & \\ & e^{-ik_z} \end{pmatrix}_\tau . \tag{5.68}$$

Here, the suffix $\tau$ meant the matrix for the lattice degrees of freedom. The Hamiltonian (5.66) satisfies

$$G_y(k_z)H_{\tilde{\nu}}^{\mathrm{LC}}(k_x,k_y,k_z)G_y(k_z)^{-1} = H_{\tilde{\nu}}^{\mathrm{LC}}(k_x,-k_y,k_z), \tag{5.69}$$

$$I(k_z)H_{\tilde{\nu}}^{\mathrm{LC}}(\mathbf{k})I(k_z)^{-1} = H_{\tilde{\nu}}^{\mathrm{LC}}(-\mathbf{k}), \tag{5.70}$$

and these operators satisfy the commutation relation

$$G_y(-k_z)I(k_z) = e^{ik_z}I(k_z)G_y(k_z), \tag{5.71}$$

which is just a Bloch form for $\hat{G}_y\hat{I} = \hat{T}_z\hat{I}\hat{G}_y$ given by the combination of Eqs. (4.20) and (4.21).

The Hamiltonian $H_{\tilde{\nu}}^{\mathrm{LC}}$ is gapped unless $m = \pm 2, 0$. The gap closes at $\Gamma$ when $m = -2$, at $Y$, $Z$ when $m = 0$, and at $C$ when $m = 2$. The glide-$Z_2$ invariant is nontrivial (trivial) if $0 < |m| < 2$ ($|m| > 2$). The bulk band structures at $m = -2.5$, $m = -2$, and $m = -1.5$ are shown in Fig. 5.4a–c. We can see the bands at $\Gamma$ have inverted between $m < -2$ (Fig. 5.4a) and $m > -2$ (Fig. 5.4c).

The effective Hamiltonian at the four high-symmetry points $P = (k_x, k_y, 0)$ ($\Gamma$, $Y$, $Z$, and $C$) in Eq. (4.28) can be written as

$$H_{\tilde{\nu}}^{\mathrm{LC}}(P) = (m + \cos k_x + \cos k_y)\sigma_z. \tag{5.72}$$

The irreducible representations (irreps) of the occupied states are summarized in Table 5.2. The sum of the numbers of the irreps $\Gamma_2^+$ at $\Gamma$, $Y_2^+$ at $Y$, $Z_2^-$ at $Z$, and $C_2^-$ at $C$ in the occupied bands is odd at $m = -1.5$ and the system is $Z_2$ nontrivial, whereas it is even at $m = -2.5$ and the system is $Z_2$ trivial.

The parities at TRIMs at $m = -1.5$ are shown in Fig. 5.3c. Therefore, it is a higher-order TI with $(\mathbb{Z}_2, \mathbb{Z}_2, \mathbb{Z}_2; \mathbb{Z}_4) = (0, 0, 0; 2)$ in ***2*** shown in Ref. [4]. Thus, the

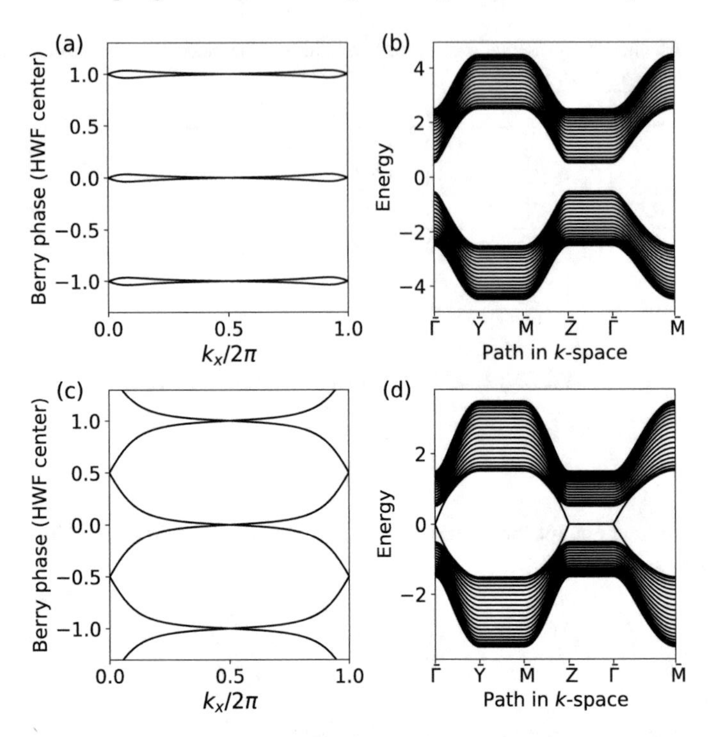


**Fig. 5.5** Hybrid Wannier centers (Berry phase) indicating bulk topologies on the $k_z = 0$ plane and surface band structures on the (100) surface from the layer construction (001; 0) of Eq. (5.66) with **a**, **b** $m = -2.5$ and **c**, **d** $m = -1.5$, respectively. The high-symmetry points are given as $\bar{\Gamma} = (0, 0)$, $\bar{Y} = (\pi, 0)$, $\bar{Z} = (0, \pi)$, and $\bar{M} = (\pi, \pi)$ on the (100) surface in **b** and **d**. **a** Gapped hybrid Wannier centers correspond to **b** the trivial surface states in the trivial phase, whereas **c** gapless hybrid Wannier centers correspond to **d** the nontrivial surface states in the glide-$Z_2$ topological phase

glide-symmetric topological phase with the nontrivial $Z_2$ invariant at $m = -1.5$ is a higher-order TI in accordance with Sect. 4.5.

Let us confirm the Berry phase and the surface states characterizing the glide-symmetric $Z_2$ magnetic TCI. We performed numerical calculations by using the open-source PythTB [5]. We numerically calculate the Berry phase (from the flow of hybrid Wannier centers) on the $k_z = 0$ plane in bulk and surface band structures for a slab geometry with (100) surfaces preserving the glide symmetry $G_y$ for the cases of $m = -2.5$ (Fig. 5.5a, b) and $m = -1.5$ (Fig. 5.5c, d), which we know are in the trivial and the glide-$Z_2$ magnetic TCI phases, respectively. $\bar{\Gamma}$, $\bar{Y}$, $\bar{Z}$, and $\bar{M}$ in Fig. 5.5b, d denote the high-symmetry points $(0, 0)$, $(\pi, 0)$, $(0, \pi)$, and $(\pi, \pi)$ on the (100) surface, respectively. In the trivial case of $m = -2.5$, the hybrid Wannier centers are solely periodic and gapped in Fig. 5.5a, and this corresponds to the trivial surface states in Fig. 5.5b. On the other hand, for the glide-symmetric $Z_2$ magnetic TCI case of $m = -1.5$, there exist gapless hybrid Wannier centers in Fig. 5.5c consistent with the existence of the gapless single surface Dirac cone protected by topology in Fig. 5.5d. The fact that the hybrid Wannier centers for two occupied states wrap the system in the opposite directions implies absence of chiral states ($C_z = 0$) but presence of the helical states ($\tilde{\nu} = +1$). Obviously, the hybrid Wannier centers on the $k_x = 0$ or $k_y = 0$ planes are always zero corresponding to $C_x = 0$ and $C_y = 0$, respectively. This totally agrees with our analytic calculations and the intuition from the schematic configuration shown in Fig. 5.3a.

We can also numerically confirm the hinge states of the higher-order TI ensured by inversion symmetry, which supports our result of Eq. (4.31). To this end, let us consider the following eight-band model. The Hamiltonian $H_{\tilde{\nu},z_4}$ consists of the Hamiltonians $H^A$ and $H^B$ describing the Chern insulator with the Chern number $+1$ and $-1$, respectively, as well as the inter-layer hopping $H^{AB}$ (Fig. 5.6a):

$$H_{\tilde{\nu},z_4} = \sum_{n,\mathbf{r}} (H_n^A + H_n^B + H_n^{AB}), \tag{5.73}$$

$$\begin{aligned} H_n^A = &\sum_{i=1,2} \left[ \psi^\dagger_{Ai,\mathbf{r}+\hat{x},n} \frac{\sigma_z + i\sigma_x}{2} \psi_{Ai,\mathbf{r},n} + \text{h.c.} \right] + \sum_{i=1,2} m \psi^\dagger_{Ai,\mathbf{r},n} \sigma_z \psi_{Ai,\mathbf{r},n} \\ &+ \left[ \psi^\dagger_{A1,\mathbf{r},n} \frac{\sigma_z + i\sigma_y}{2} \psi_{A2,\mathbf{r},n} + \psi^\dagger_{A1,\mathbf{r}+\hat{y},n} \frac{\sigma_z + i\sigma_y}{2} \psi_{A2,\mathbf{r},n} + \text{h.c.} \right], \end{aligned} \tag{5.74}$$

$$\begin{aligned} H_n^B = &\sum_{i=1,2} \left[ \psi^\dagger_{Bi,\mathbf{r}+\hat{x},n} \frac{\sigma_z + i\sigma_x}{2} \psi_{Bi,\mathbf{r},n} + \text{h.c.} \right] + \sum_{i=1,2} m \psi^\dagger_{Bi,\mathbf{r},n} \sigma_z \psi_{Bi,\mathbf{r},n} \\ &+ \left[ \psi^\dagger_{B1,\mathbf{r},n} \frac{\sigma_z - i\sigma_y}{2} \psi_{B2,\mathbf{r},n} + \psi^\dagger_{B1,\mathbf{r}+\hat{y},n} \frac{\sigma_z - i\sigma_y}{2} \psi_{B2,\mathbf{r},n} + \text{h.c.} \right], \end{aligned} \tag{5.75}$$

$$H_n^{AB} = \frac{1}{2} \left[ \psi^\dagger_{A1,\mathbf{r},n} \psi_{B1,\mathbf{r},n} + \psi^\dagger_{A2,\mathbf{r},n+1} \psi_{B2,\mathbf{r},n} + \text{h.c.} \right], \tag{5.76}$$

where $\mathbf{r}$ runs over the lattice sites within the $xy$ plane $\mathbf{r}_i = (r_1, r_2), r_1 \in \mathbb{Z}, r_2 \in \mathbb{Z} \pm \frac{1}{4}$. When the term $H^{AB}$ is absent, it is nothing other than the model of the LC with (001;0) shown in Fig. 5.3a. Because this configuration exhibits the surface states, we add the term $H^{AB}$ to see the hinge states separated from the surface states. We note that this model is invariant under glide symmetry and inversion symmetry.

The Hamiltonian $H_{\tilde{\nu},z_4}$ shows the glide-$Z_2$ magnetic TCI phase if $-1.5 < m < -0.5$ and $0.5 < m < 1.5$, the Weyl semimetal phase if $1.5 < |m| < 2.5$ and $|m| < 0.5$, and a trivial phase if $2.5 < |m|$. Let us focus on the glide-$Z_2$ magnetic TCI phase at $m = -1$. Figure 5.6b–d are band structures of a slab with (b) the (100) surface, (c) the (010) surface, and (d) the (001) surface. The gapless surface states emerge on the (100) surface preserving the glide symmetry [Fig. 5.6b] while the band structures are gapped on the (010) surface [Fig. 5.6c] and the (001) surface [Fig. 5.6d]. Now we consider rod configurations. To separate the hinge states away from the surface states, we make a rod configuration having a periodicity along $x$ direction. The rod band structure exhibits the gapless states depicted in Fig. 5.6e. The wavefunctions of the gapless states are strongly localized on hinges of the rod which is supported by Fig. 5.6f representing the density of the wavefunctions of the

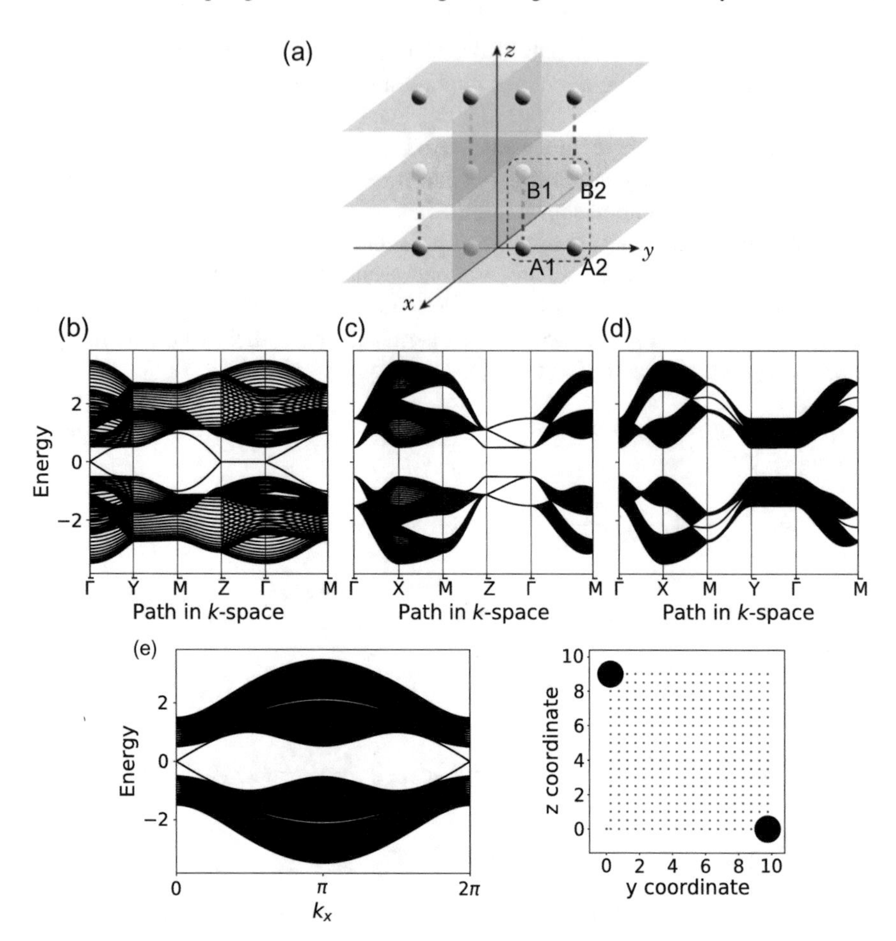


**Fig. 5.6** **a** Schematic figure of our model showing a glide-$Z_2$ topological phase with four sites in the unit cell marked by the black dotted line. If the inter-layer hoppings (red dotted lines) are absent, it is nothing other than the layer construction (001;0) shown in Fig. 5.3a. The model is manifestly invariant under glide symmetry and inversion symmetry. **b-d** Band structures for slabs with **b** the (100) surfaces, **c** the (010) surfaces, and **d** the (001) surfaces. The thickness of the slab is $n = 20$. In **b**, the gapless surface states exist on the (100) surface which preserves glide symmetry. On the other hand, the surface states are gapped on **c** the (010) surface and **d** the (001) surface breaking glide symmetry. The high-symmetry points are given as $\bar{\Gamma} = (0, 0)$, $\bar{M} = (\pi, \pi)$ on the (100), (010), and (001) surfaces, $\bar{Y} = (\pi, 0)$, $\bar{Z} = (0, \pi)$ on the (100) surface, $\bar{X} = (\pi, 0)$, $\bar{Z} = (0, \pi)$ on the (010) surface, and $\bar{X} = (\pi, 0)$, $\bar{Y} = (0, \pi)$ on the (001) surface. **e** The gapless band structure for the rod configuration. The rod is finite along the $y$ and $z$ direction with the thickness equal to $n = 10$, and is periodic along the $x$ direction. It is gapless, with the gapless states lying on hinges. **f** Density plot of the wavefunctions of the gapless states. They are localized on hinges of the rod

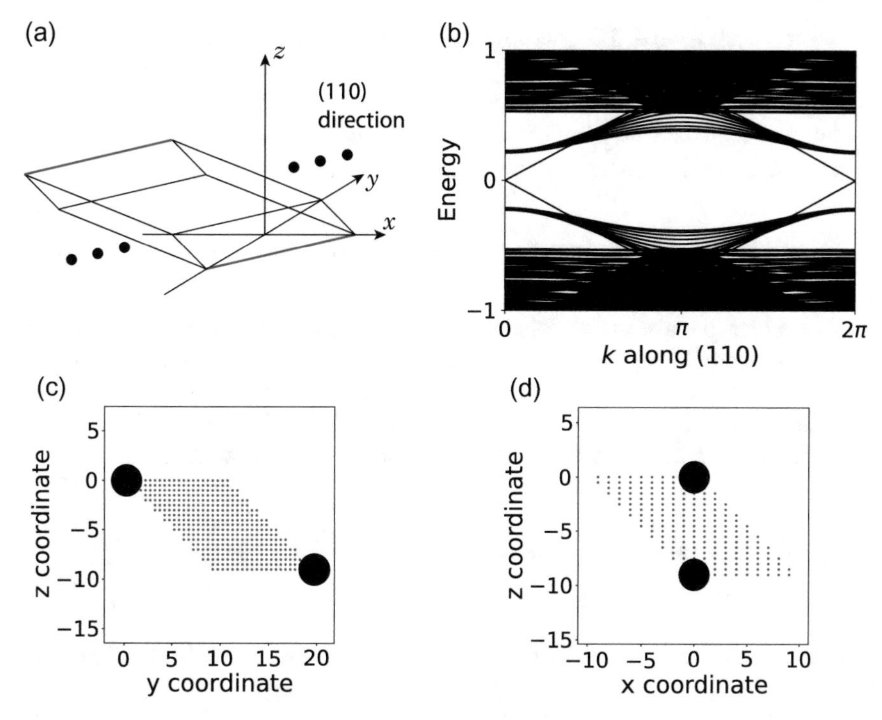


**Fig. 5.7** **a** An arbitrary rod configuration with a periodicity along the (110) direction and surfaces whose Miller indices are $(\bar{1}10)$ and (001). The thickness of the rod is $n = 10$. **b** The band structure for the rod configuration. The gapless hinge states emerge within the gap of the surface states. Density plots of the wavefunctions of the gapless states **c** on the $yz$ plane and **d** the $xz$ plane. They localize on hinges of the rod denoting red lines in **a**

gapless states. Therefore, these results support the relation $\bar{\nu} = z_4/2$ in the previous chapter.

We also calculate the band structures with an arbitrary rod structure in Fig. 5.7a. This rod configuration has a periodicity along the (110) direction and those surfaces are given by Miller indices $(\bar{1}10)$ and (001). The rod band structure is shown in Fig. 5.7b. One can see the gapless states between the gapped surface states distinguished from the bulk states. The density of the wavefunctions of the gapless states on the $yz$ plane and the $xz$ plane is shown in Fig. 5.7c, d, respectively. Therefore, the gapless states are localized on hinges of a rod (Fig. 5.7a), and the relation $\bar{\nu} = z_4/2$ is confirmed.

#### 5.2.1.2 Chern Insulator Phase: (010; $\frac{1}{2}$)

Next, we consider the eLC (010; $\frac{1}{2}$) showing a weak 3D Chern insulator with a nontrivial Chern number along the $xz$ direction (Fig. 5.3b). In the absence of inversion symmetry, such a weak 3D Chern insulator phase can be realized as a stacking of 2D Chern insulators with $C_y = +1$ parallel to the $xz$ plane, located at $y = y_0 + n$ with an integer $n$. When we add inversion symmetry with its inversion center at the origin, $y_0$ is set to be 0 or $\frac{1}{2}$ (Fig. 5.3b). To make $\tilde{\nu}$ to vanish, we choose $y_0 = \frac{1}{2}$. This layer is described as $(010; \frac{1}{2})$. Therefore, we have a representative Hamiltonian

$$H_C^{\mathrm{LC}}(\mathbf{k}) = \left(m + \cos k_x + \cos \frac{k_z}{2}\right)\sigma_z + \sin k_x \sigma_y + \sin \frac{k_z}{2}\sigma_x. \tag{5.77}$$

Here, in order to make the model consistent with the glide symmetry, the primitive translation vector along the $z$ axis is taken as $(0, 0, 1/2)$. Nonetheless, for the calculation of topological invariants from Eqs. (4.28) and (4.32) we should instead regard the primitive translation vector to be $(0, 0, 1)$; namely we double the unit cell along the $z$ direction. In this doubled unit cell, the Hamiltonian is rewritten as

$$\tilde{H}_C^{\mathrm{LC}}(\mathbf{k}) = [(m + \cos k_x)\sigma_z + \sin k_x \sigma_y] + \left(\cos \frac{k_z}{2}\sigma_z + \sin \frac{k_z}{2}\sigma_x\right)\left(\cos \frac{k_z}{2}\tau_x + \sin \frac{k_z}{2}\tau_y\right). \tag{5.78}$$

The Hamiltonian $\tilde{H}_C^{\mathrm{LC}}$ indeed satisfies the relations in Eqs. (5.69) and (5.70) under the corresponding operators given by Eqs. (5.67) and (5.68).

This model shows phase transitions at $m = 0, \pm 2$. For example, we focus on the transition at $m = -2$. It is a trivial insulator at $m = -2.5$ and a weak 3D Chern insulator with $C_y = +1$ at $m = -1.5$, as can be seen from Eq. (5.77). Meanwhile, both for $m = -2.5$ and $m = -1.5$, the glide-$Z_2$ invariant (4.28) is trivial from the irreps in Table 5.2. Therefore, the Hamiltonian $\tilde{H}_C^{\mathrm{LC}}$ realizes a topological phase with a nonzero Chern number but a trivial glide-$Z_2$ invariant at $m = -1.5$ as we intuitively expected. The parities at high-symmetry points in this model are shown in Fig. 5.3d. It gives the values of the symmetry-based indicators as $(\mathbb{Z}_2, \mathbb{Z}_2, \mathbb{Z}_2; \mathbb{Z}_4) = (0, 1, 0; 0)$.

These results are also confirmed by numerical calculations. Figure 5.4d–f show the bulk band structure for $\tilde{H}_C^{\mathrm{LC}}$ at (d) $m = -2.5$, (e) $m = -2$, and (f) $m = -1.5$. One can find the second and third bands at $\Gamma$ are inverted between $m < -2$ (Fig. 5.4d) and $m > -2$ (Fig. 5.4f). The Berry phase (from the flow of hybrid Wannier centers) on the $k_y = 0$ plane in bulk and surface band structures for a slab geometry with (100) are shown in Fig. 5.8. In Fig. 5.8b, d, $\bar{\Gamma}$, $\bar{Y}$, $\bar{Z}$, and $\bar{M}$ refer to the high-symmetry points $(0, 0)$, $(\pi, 0)$, $(0, \pi)$, and $(\pi, \pi)$ on the (100) surface, respectively. In the trivial case of $m = -2.5$, the hybrid Wannier centers are gapped in Fig. 5.8a and there are no gapless surface states in Fig. 5.8b. For the Chern insulator case of $m = -1.5$, the hybrid Wannier centers shift up by one unit as $k_z$ cycles by $4\pi$ in Fig. 5.8c, corresponding to the Chern number $C_y$ equal to one.[3] Therefore, we can confirm the gapless chiral surface states in Fig. 5.8d. Obviously, the hybrid Wannier centers on the $k_x = 0$ or $k_z = 0$ planes are always zero corresponding to $C_x = 0$ and $C_z = 0$,

[3] Because of the branch cut, the hybrid Wannier center for a single band has to have a period of $4\pi$ for $k_z$ on the glide-invariant plane $k_y = 0$.

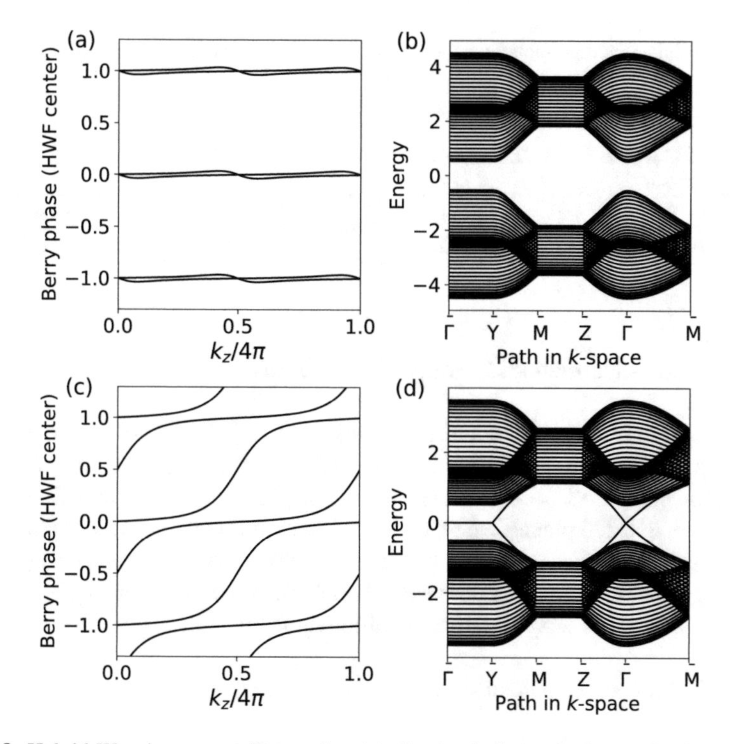


**Fig. 5.8** Hybrid Wannier centers (Berry phase) indicating bulk topologies on the $k_y = 0$ plane and surface band structures on the (100) surface from the layer construction (010; $\frac{1}{2}$) of Eq. (5.78) with **a**, **b** $m = -2.5$ and **c**, **d** $m = -1.5$, respectively. The high-symmetry points are given as $\bar{\Gamma} = (0, 0)$, $\bar{Y} = (\pi, 0)$, $\bar{Z} = (0, \pi)$, and $\bar{M} = (\pi, \pi)$ on the (100) surface in **b** and **d**. **a** Gapped hybrid Wannier centers correspond to **b** the trivial surface states in the trivial phase, whereas **c** gapless hybrid Wannier centers correspond to **d** the nontrivial surface states in the Chern insulator phase. Due to the branch cut, the periodicity of a single band should be $4\pi$ along $k_z$ on the glide-invariant plane $k_y = 0$

respectively, and the glide-$Z_2$ invariant $\tilde{\nu}$ is also zero, which is exactly consistent with our analytic calculations and the intuition from the schematic configuration shown in Fig. 5.3b.

### 5.2.2 Tight-Binding Models Constructed from the Layer Construction for *14*

Next, we study the two minimal LCs (001; 0) and (020; 0) in Table 5.1 for ***14***. These two LCs correspond to the glide-symmetric $Z_2$ magnetic TCI and the Chern insulator phases in ***14***, and in this subsection we construct simple tight-binding models for them in order to examine our scenario.

**Table 5.3** Irreducible representations for the two tight-binding models for the topological phases in ***14*** from the layer construction

| | $\Gamma$ | $Y$ | $D$ | $E$ |
|---|---|---|---|---|
| $H^{\mathrm{LC}}_{\tilde{\nu}}(m=-2.5)$ | $\Gamma_1^+ + \Gamma_2^+$ | $Y_1^+ + Y_2^+$ | $D_1^+ + D_2^+$ | $E_1^+ + E_2^+$ |
| $H^{\mathrm{LC}}_{\tilde{\nu}}(m=-1.5)$ | $\Gamma_1^- + \Gamma_2^-$ | $Y_1^+ + Y_2^+$ | $D_1^+ + D_2^+$ | $E_1^+ + E_2^+$ |
| $H^{\mathrm{LC}}_{C}(m=-2.5)$ | $\Gamma_1^+ + \Gamma_2^+$ | $Y_1^+ + Y_2^+$ | $D_1^+ + D_2^+$ | $E_1^+ + E_2^+$ |
| $H^{\mathrm{LC}}_{C}(m=-1.5)$ | $\Gamma_1^- + \Gamma_2^-$ | $Y_1^+ + Y_2^+$ | $D_1^+ + D_2^+$ | $E_1^+ + E_2^+$ |

#### 5.2.2.1 Glide-Symmetric $Z_2$ Magnetic TCI Phase: (001; 0)

We first calculate the topological invariants for the LC $(001; 0)$. A Hamiltonian exhibiting a glide-symmetric $Z_2$ magnetic TCI is constructed by putting 2D Chern insulator layers with $C_z = +1$ on the $z = z_0 + n$ planes and those with $C_z = -1$ on the $z = z_0 + n + 1/2$ planes, where $C_z$ is the Chern number along the $xy$ plane and $n$ is an integer. Since this configuration becomes compatible with inversion symmetry by setting $z_0 = 0$, the layer is now described as $(001; 0)$ as shown in Fig. 5.9a. By using Eq. (2.118), its representative Hamiltonian is given by

$$H^{\mathrm{LC}}_{\tilde{\nu}}(\mathbf{k}) = (m + \cos k_x + \cos k_y)\sigma_z + \sin k_x \sigma_x + \sin k_y \sigma_y \tau_z. \tag{5.79}$$

The form of the Hamiltonian is the same as Eq. (5.66), whereas the corresponding glide and inversion operators in momentum space are taken as

$$G_y(k_z) = e^{-ik_z/2}\left(\cos\frac{k_z}{2}\tau_x + \sin\frac{k_z}{2}\tau_y\right), \tag{5.80}$$

$$I(k_y, k_z) = \sigma_z \begin{pmatrix} 1 & \\ & e^{-i(k_y+k_z)} \end{pmatrix}_\tau . \tag{5.81}$$

Thus, we can easily see

$$G_y(-k_z) I(k_y, k_z) = e^{i(-k_y+k_z)} I(-k_y, k_z) G_y(k_z), \tag{5.82}$$

which is Bloch form of $\hat{G}_y \hat{I} = \hat{T}_y^{-1}\hat{T}_z \hat{I} \hat{G}_y$.

As we mentioned in the previous section, the representative Hamiltonian $H^{\mathrm{LC}}_{\tilde{\nu}}(\mathbf{k})$ in Eq. (5.79) closes the gap at $\Gamma$ when $m = -2$, at $Y, Z$ when $m = 0$, and at $C$ when $m = 2$. The irreps for the phases at $m = -2.5$ and $m = -1.5$ are summarized in Table 5.3. By calculating the $C_2$ screw eigenvalues at $\Gamma$, $D$, $Y$, and $E$, the value of the symmetry-based indicator in our new formula of Eq. (4.44) is $-1$, i.e., it is nontrivial. Since the Chern number should be 0 in this configuration, this nontrivial indicator is attributed to a nontrivial glide-$Z_2$ invariant. Since this $Z_2$ topological phase corresponds to $(\mathbb{Z}_2, \mathbb{Z}_2, \mathbb{Z}_2; \mathbb{Z}_4) = (0, 0, 0; 2)$ in **2** by calculating the parities (Fig. 5.9c), this model is also a higher-order TI ensured by inversion symmetry.

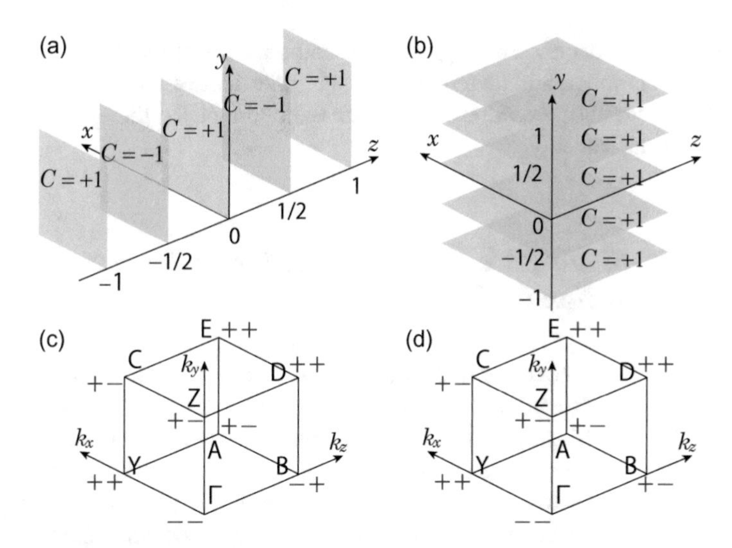


**Fig. 5.9** Layer constructions for **a** a glide-$Z_2$ topological phase and **b** a Chern insulator with ***14***. The parities at the high-symmetry points for the models **a** and **b** are shown in **c** and **d**, respectively. Unlike the case of ***13***, the combinations of parities at high-symmetry points are equivalent between **c** and **d**

### 5.2.2.2 Chern Insulator Phase: (020; 0)

Next we construct a model in the Chern insulator phase. To make a model compatible with inversion symmetry, we must put *two* 2D Chern insulator layers with $C_y = +1$ within the lattice constant, i.e., along the $xz$ plane at $y = n$ and $y = n + 1/2$ ($n$ : integer) (Fig. 5.9b). It is written as (020; 0). One can write down the model as

$$H_C^{\mathrm{LC}}(\mathbf{k}) = [(m + \cos k_x + \cos k_z)\sigma_z + \sin k_x\sigma_y + \sin k_z\sigma_x]\tau_0 \tag{5.83}$$

with the glide operator and inversion operator given by Eqs. (5.80) and (5.81), respectively. This model is simply stacked by *two* 2D Chern insulator in Eq. (2.118). Therefore, we can easily see $H_C^{\mathrm{LC}}$ closes its gap at $m = 0, \pm 2$, and phase transitions occur. We here focus on the transition at $m = -2$, when the gap closes at $\Gamma$, and the irreps at the both sides of the gap closing are given in Table 5.3. Surprisingly, they are the same as the previous model with the Hamiltonian $H_{\tilde{\nu}}^{\mathrm{LC}}$ for the glide-symmetric TCI ($\tilde{\nu} = 1, C_y = 0$) despite of the difference in the models. Because the Chern number for $H_C^{\mathrm{LC}}$ is obviously equal to two by construction, the glide-$Z_2$ invariant is zero in this model. Thus, the Chern insulator with the Chern number equal to $4n + 2$ ($n$: integer) is also a higher-order TI ensured by inversion symmetry which has $(\mathbb{Z}_2, \mathbb{Z}_2, \mathbb{Z}_2; \mathbb{Z}_4) = (0, 0, 0; 2)$ in **2** (Fig. 5.9d).

The two models (5.79) and (5.83) give the same values for the symmetry-based indicator. This is natural from our result in (4.44). From Eq. (4.44), the symmetry-based indicator is given by a combination of the glide-$Z_2$ invariant and the Chern number, and therefore, although the values of the glide-$Z_2$ invariant and the Chern number are different between the two models, the symmetry-based indicators are the same.

### *5.2.3 Tight-Binding Models Constructed from the Layer Construction for 15*

Finally, we study the two minimal LCs (001; 0) and ($\bar{1}10$; 0) in Table 5.1 for ***15***. These two LCs correspond to the glide-symmetric $Z_2$ magnetic TCI and the Chern insulator phases in ***15***, respectively, and we construct simple tight-binding models for them in order to investigate our scenario in this subsection. The main difference from the previous space groups ***13*** and ***14*** is the primitive translation vectors. We show the outline of the calculations, and the details are presented in Sect. 5.3.

#### 5.2.3.1 Glide-Symmetric $Z_2$ Magnetic TCI Phase: (001; 0)

Let us begin with the glide-symmetric $Z_2$ magnetic TCI phase. The glide-symmetric $Z_2$ magnetic TCI phase can be realized as an alternate stacking of two 2D Chern insulator layers with opposite signs of the Chern number $\pm 1$. Suppose layers of a 2D Chern insulator with $C_z = \pm 1$ along the $xy$ plane are placed at $z = z_0 + n$ in which $n$ is an integer in the absence of inversion symmetry, i.e., ***9***. The glide symmetry then requires presence of other layers of a 2D Chern insulator with $C_z = -1$ at $z = z_0 + n + 1/2$, because the glide operation changes the sign of the Chern number. This model gives a nontrivial $Z_2$ topological invariant $\tilde{\nu}$ for glide-symmetric systems by a direct calculation.

We now consider the case when inversion symmetry is added. Inversion symmetry fixes the value of $z_0$ to be $z_0 = 0$, and the configuration allowed by symmetry is the one with the 2D Chern insulators on $z = n$ with $C_z = +1$ and on $z = n + 1/2$ with $C_z = -1$, where $n$ is an integer (Fig. 5.10a). Therefore, a representative Hamiltonian for the glide-symmetric $Z_2$ magnetic TCI phase with additional inversion symmetry is obtained as[4]

$$H_{\tilde{\nu}}^{\mathrm{LC}} = \left(m + 2\cos\frac{k_x}{2}\cos\frac{k_y}{2}\right)\sigma_z + \sin\frac{k_x}{2}\cos\frac{k_y}{2}(\sigma_x + \sigma_y) + \cos\frac{k_x}{2}\sin\frac{k_y}{2}(\sigma_x - \sigma_y)\tau_z \tag{5.84}$$

by using Eq. (2.118).

[4] Details are given in Sect. 5.3.

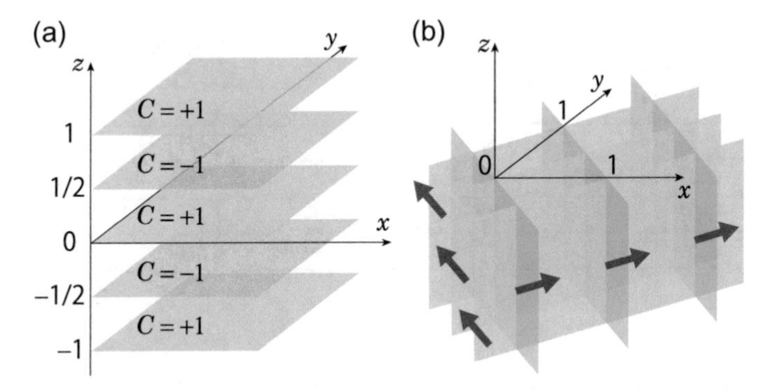


**Fig. 5.10** Layer constructions for **a** (001; 0), showing a glide-$Z_2$ topological phase ($\tilde{\nu} = 1, C_y = 0$) and **b** ($\bar{1}10$; 0), showing a Chern insulator ($\tilde{\nu} = 0, C_y = 2$) with ***15***. The red arrows denote the directions of the normal vector of the plane giving a positive Chern number +1

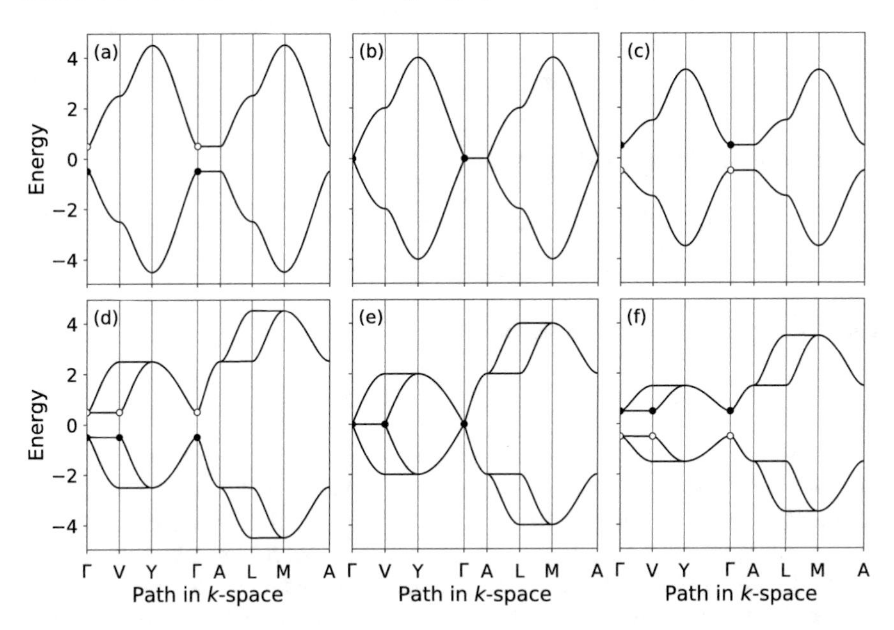


**Fig. 5.11** Bulk band structures for the tight-binding models for **a–c** the glide-symmetric $Z_2$ topological crystalline insulator phase of Eq. (5.84) and **d–e** the Chern insulator phase of Eq. (5.92) constructed from the layer construction in ***15*** with **a**, **d** $m = -2.5$; **b**, **e** $m = -2$; **c**, **f** $m = -1.5$. Filled (open) circles mark states of common character; band inversions at $\Gamma$ and $V$ are evident in **c** and **f**

**Table 5.4** Irreducible representations for the two tight-binding models for the topological phases in ***15*** from the layer construction

| | $\Gamma$ | $Y$ | $V$ | $M$ | $L$ |
|---|---|---|---|---|---|
| $H_{\tilde{\nu}}^{\mathrm{LC}}(m=-2.5)$ | $\Gamma_1^+ + \Gamma_2^+$ | $Y_1^+ + Y_2^+$ | $2V_1^+$ | $M_1$ | $L_1^+ + L_1^-$ |
| $H_{\tilde{\nu}}^{\mathrm{LC}}(m=-1.5)$ | $\Gamma_1^- + \Gamma_2^-$ | $Y_1^+ + Y_2^+$ | $2V_1^+$ | $M_1$ | $L_1^+ + L_1^-$ |
| $H_C^{\mathrm{LC}}(m=-2.5)$ | $\Gamma_1^+ + \Gamma_2^+$ | $Y_1^+ + Y_2^+$ | $2V_1^+$ | $M_1$ | $L_1^+ + L_1^-$ |
| $H_C^{\mathrm{LC}}(m=-1.5)$ | $\Gamma_1^- + \Gamma_2^-$ | $Y_1^+ + Y_2^+$ | $V_1^+ + V_1^-$ | $M_1$ | $L_1^+ + L_1^-$ |

The corresponding $k$-dependent glide operator and $k$-dependent inversion operator are

$$G_y(k_z) = e^{-ik_z/2}\left(\cos\frac{k_z}{2}\tau_x + \sin\frac{k_z}{2}\tau_y\right), \tag{5.85}$$

$$I(k_z) = \sigma_z \begin{pmatrix} 1 & 0 \\ 0 & e^{-ik_z} \end{pmatrix}_\tau . \tag{5.86}$$

The Hamiltonian in Eq. (5.84) satisfies

$$G_y(k_z)H_{\tilde{\nu}}^{\mathrm{LC}}(k_x,k_y,k_z)G_y(k_z)^{-1} = H_{\tilde{\nu}}^{\mathrm{LC}}(k_x,-k_y,k_z), \tag{5.87}$$

$$I(k_z)H_{\tilde{\nu}}^{\mathrm{LC}}(\mathbf{k})I(k_z)^{-1} = H_{\tilde{\nu}}^{\mathrm{LC}}(-\mathbf{k}), \tag{5.88}$$

and those operators satisfy the algebra

$$G_y(k_z)I(k_z) = e^{ik_z}I(k_z)G_y(k_z). \tag{5.89}$$

The Hamiltonian $H_{\tilde{\nu}}^{\mathrm{LC}}$ is gapped unless $m = \pm 2, 0$. The gap closes at $\Gamma$ when $m=-2$, at $V$ (and $V'$) when $m=0$, and at $Y$ when $m=2$. The $Z_2$ topological invariant is nontrivial (trivial) if $0<|m|<2$ ($|m|>2$). The bulk band structures at $m=-2.5$, $m=-2$, and $m=-1.5$ are shown in Fig. 5.11a–c. We can find the bands at $\Gamma$ are inverted between $m<-2$ (Fig. 5.11a) and $m>-2$ (Fig. 5.11c).

The effective Hamiltonian at the high-symmetry points $P=(k_x,k_y,0)$ ($\Gamma$, $Y$, and $V$) in Eq. (4.81) can be written as

$$H_{\tilde{\nu}}^{\mathrm{LC}}(P) = \left(m + 2\cos\frac{k_x}{2}\cos\frac{k_y}{2}\right)\sigma_z. \tag{5.90}$$

The irreps of the occupied states are summarized in Table 5.4. Thus, it is $Z_2$ nontrivial when $|m|<2$, and trivial otherwise.

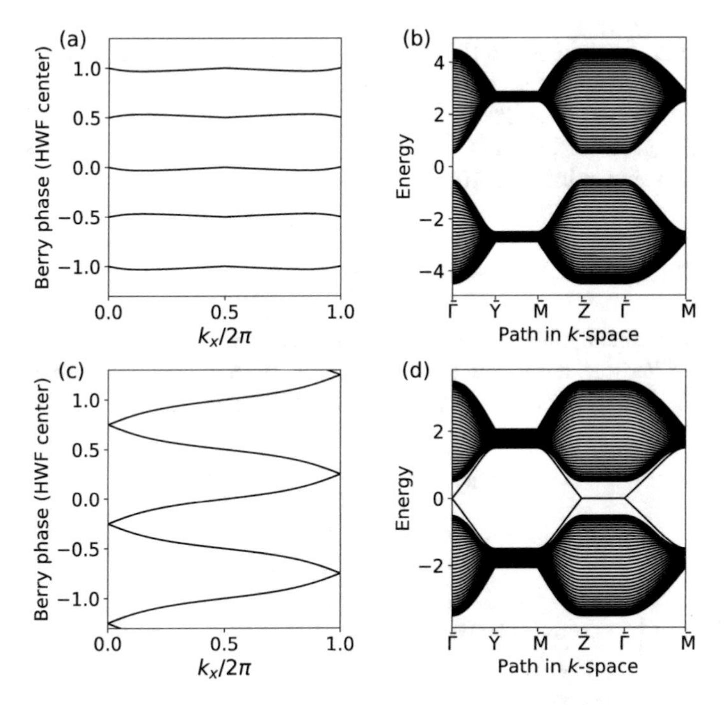


**Fig. 5.12** Hybrid Wannier centers (Berry phase) indicating bulk topologies on the $k_z = 0$ plane and surface band structures on the (100) surface from the the layer construction (001; 0) of Eq. (5.84) in ***15*** with **a**, **b** $m = -2.5$ and **c**, **d** $m = -1.5$, respectively. The high-symmetry points are given as $\bar{\Gamma} = (0, 0)$, $\bar{Y} = (\pi, 0)$, $\bar{Z} = (0, \pi)$, and $\bar{M} = (\pi, \pi)$ on the (100) surface in **b** and **d**. **a** Gapped hybrid Wannier centers correspond to **b** the trivial surface states in the trivial phase, whereas **c** gapless hybrid Wannier centers correspond to **d** the nontrivial surface states in the glide-$Z_2$ topological phase

The $z_4$ indicator at $m = -1.5$ is equal to *two* from Table 5.4. Therefore, it is a higher-order TI with $(\mathbb{Z}_2, \mathbb{Z}_2, \mathbb{Z}_2, \mathbb{Z}_4) = (0, 0, 0; 2)$ in ***2***. Thus, the glide-symmetric topological phase with the nontrivial $Z_2$ topological invariant at $m = -1.5$ is a higher-order TI in accordance with Sect. 4.5.

Let us perform numerical calculations in order to show behaviors of the Berry phase and the surface states enriching the glide-symmetric $Z_2$ magnetic TCI. The results are shown in Fig. 5.12. We calculate the Berry phase (from the flow of hybrid Wannier centers) on the $k_z = 0$ plane in bulk and surface band structures for a slab geometry with (100) surfaces for the cases of $m = -2.5$ (Fig. 5.12a, b) and $m = -1.5$ (Fig. 5.12c, d), which we know are in the trivial and the glide-$Z_2$ magnetic TCI phases, respectively. $\bar{\Gamma}$, $\bar{Y}$, $\bar{Z}$, and $\bar{M}$ in Fig. 5.12b, d denote $(0, 0)$, $(\pi, 0)$, $(0, \pi)$, and $(\pi, \pi)$ on the (100) surface, respectively. In the trivial case of $m = -2.5$, the hybrid Wannier centers are solely periodic and gapped in Fig. 5.12a, consistent with the absence of topological surface states in Fig. 5.12b. On the other hand, for the glide-symmetric $Z_2$ magnetic TCI case of $m = -1.5$, the hybrid Wannier centers are gapless in Fig. 5.12c, consistent with the existence of the gapless single surface Dirac cone protected by topology in Fig. 5.12d. Obviously, the hybrid Wannier centers on the $k_x = 0$ or $k_y = 0$ planes are always zero corresponding to $C_x = 0$ and $C_y = 0$, respectively, and the total Chern number on the $k_z = 0$ plane is also zero ($C_z = 0$). This exactly agrees with our analyses and the intuitive expectation from the schematic configuration shown in Fig. 5.10a.

#### 5.2.3.2 Chern Insulator Phase: $(\bar{1}10; 0)$

Next, we construct a model with a nontrivial Chern number. In this case, we simply stack a 2D Chern insulator with $C = +1$ along $(\bar{1}10)$ and $(110)$ directions by using Eq. (2.118). Then, we have a representative Hamiltonian

$$H_C^{\mathrm{LC}}(\mathbf{k}) = \begin{pmatrix} H^{(+)}(\mathbf{k}) & \\ & H^{(-)}(\mathbf{k}) \end{pmatrix}, \tag{5.91}$$

where

$$H^{(\pm)}(\mathbf{k}) = \left(m + \cos k_z + \cos \frac{k_x \pm k_y}{2}\right)\sigma_z + \sin k_z \sigma_x \pm \sin \frac{k_x \pm k_y}{2}\sigma_y. \tag{5.92}$$

The Hamiltonian $\tilde{H}_C^{\mathrm{LC}}$ indeed satisfies the relations in Eqs. (5.87) and (5.88) under the corresponding operators given by Eqs. (5.85) and (5.86).

This model shows phase transitions at $m = 0, \pm 2$. For example, we focus on the transition at $m = -2$. It is a trivial insulator at $m = -2.5$ and a Chern insulator with $C_y = +2$ at $m = -1.5$, as can be seen from Eq. (5.92). Meanwhile, both $m = -2.5$ and $m = -1.5$, the $Z_2$ topological invariant (4.81) is trivial. Therefore, the Hamiltonian $\tilde{H}_C^{\mathrm{LC}}$ realizes a topological phase with nonzero Chern number but a trivial $Z_2$ topological invariant at $m = -1.5$ as we have expected.

These results are also confirmed by numerical calculations. Figure 5.11d–f show the bulk band structure for $\tilde{H}_C^{\mathrm{LC}}$ at (d) $m = -2.5$, (e) $m = -2$, and (f) $m = -1.5$. One can find the bands at $\Gamma$ and the second and third bands at $V$ have inverted between $m < -2$ (Fig. 5.11d) and $m > -2$ (Fig. 5.11f). The Berry phase (from the flow of hybrid Wannier centers) on the $k_y = 0$ plane in bulk and surface band structures for a slab geometry with (100) are shown in Fig. 5.13a, b for $m = -2.5$ and (c) and (d) for $m = -1.5$. In Fig. 5.13b, d, $\bar{\Gamma}$, $\bar{Y}$, $\bar{Z}$, and $\bar{M}$ refer to the high-symmetry points $(0, 0)$, $(\pi, 0)$, $(0, \pi)$, and $(\pi, \pi)$ on the (100) surface, respectively. In the trivial case of $m = -2.5$, the hybrid Wannier centers are gapped in Fig. 5.13a and there are no gapless surface states in Fig. 5.13b. For the Chern insulator case of $m = -1.5$, the doubly-degenerated hybrid Wannier centers shift up by one unit as $k_z$ cycles by $4\pi$ in Fig. 5.13c corresponding to the Chern number $C_y$ to be two. Obviously, both of the Chern numbers along $(\bar{1}10)$ and $(110)$ are found to be one. Therefore, we can find the doubly-degenerated gapless chiral surface states in Fig. 5.13d. The hybrid Wannier centers on the $k_x = 0$ or $k_z = 0$ planes are obviously always zero corresponding to $C_x = 0$ and $C_z = 0$, respectively, and the glide-$Z_2$ invariant $\tilde{\nu}$ is also zero. This is

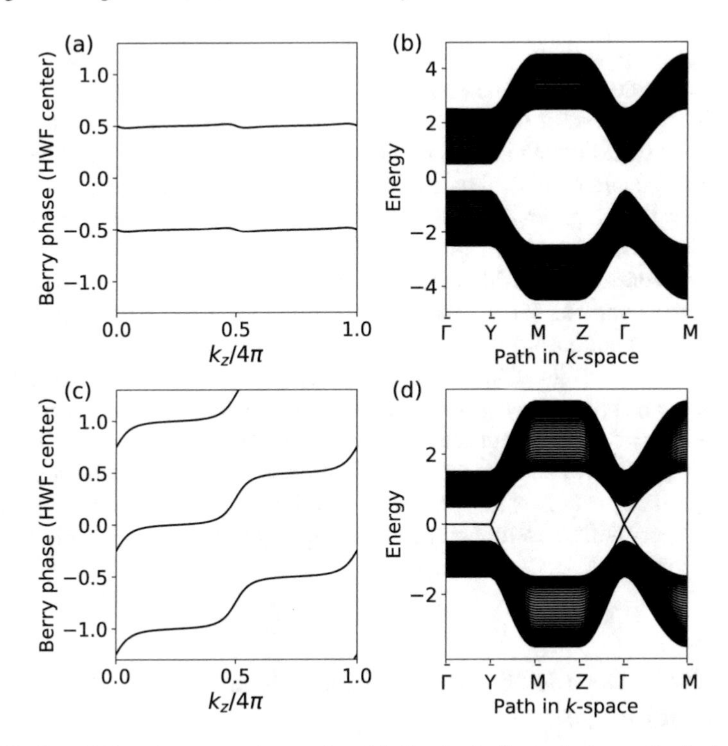


**Fig. 5.13** Hybrid Wannier centers (Berry phase) indicating bulk topologies on the $k_y = 0$ plane and surface band structures on the (100) surface from the layer construction $(\bar{1}10; 0)$ of Eq. (5.92) in ***15*** with **a**, **b** $m = -2.5$ and **c**, **d** $m = -1.5$, respectively. The high-symmetry points are given as $\bar{\Gamma} = (0, 0)$, $\bar{Y} = (\pi, 0)$, $\bar{Z} = (0, \pi)$, and $\bar{M} = (\pi, \pi)$ on the (100) surface in **b** and **d**. **a** Gapped hybrid Wannier centers correspond to **b** the trivial surface states in the trivial phase, whereas **c** gapless hybrid Wannier centers correspond to **d** the nontrivial surface states in the Chern insulator phase. Due to the branch cut, the periodicity of a single band should be $4\pi$ along $k_z$ on the glide-invariant plane $k_y = 0$

totally consistent with our analyses and the intuitive expectation from the schematic configuration shown in Fig. 5.10b.

## 5.3 Conclusion and Discussion

We have studied the layer construction for the glide-symmetric systems with and without inversion symmetry for ***7***, ***9***, ***13***, ***14***, and ***15*** when time-reversal symmetry is not enforced in the present chapter. As an independent approach from the $k$-space topology, we constructed invariants for layer constructions in these space groups only from real-space geometry of layers. This list of invariants for layer constructions turns out to agree with the topological invariants discussed in the previous chapter constructed in the $k$-space topology.

We also found the list of elementary layer constructions for these space groups, and construct tight-binding models for elementary layer constructions in ***13***, ***14***, and ***15***. By analytic and numerical calculations, we can show that they have nontrivial values either for the glide-$Z_2$ topological invariant or for the Chern number, which confirms the conclusions in the previous and the present chapters.

The layer constructions without time-reversal symmetry are classified in Ref. [4]. Nevertheless, invariants for topological crystalline insulators are not defined, and therefore their relationship with layer construction is not clarified. On the other hand, our definition of invariants for topological crystalline insulators and their direct evaluations can apply to any space group if one can define a mirror Chern invariant and $n$-fold ($n = 3, 4, 6$) rotation/screw invariants. Hence, our theory in this thesis gives a comprehensive understanding of topological phases in the absence of time-reversal symmetry in both of the approaches of the real-space and $k$-space topologies. Thus, our theory will be useful for understanding and realizing a number of magnetic topological materials.

## Appendix 1: Details of Construction of Models for Topological Phases

In this section we show a detailed construction of the models for the glide-symmetric TCI phase and the Chern insulator phase in ***13*** and ***15***. The symmetry generators are given by

$$\hat{G}_y = \left\{ M_y \middle| \frac{1}{2}\hat{\mathbf{z}} \right\}, \quad \hat{I} = \{I|\mathbf{0}\}. \tag{5.93}$$

### *1.1: Derivation of Tight-Binding Models for 13*

#### 1.1.1 The Layer Construction (001; 0) ($\tilde{\nu} = 1, C_y = 0$)

The model Hamiltonian is constructed by putting a Chern insulator with $C_z = +1$ on the $xy$ plane at $z = 0$ with inversion symmetry

$$\hat{I}\psi^\dagger(x, y)\hat{I}^{-1} = \psi^\dagger(-x, -y)\sigma_z, \tag{5.94}$$

given by Eq. (2.118), and by making copies by the glide transformation

$$\hat{G}_y \psi^\dagger(x, y, z) \hat{G}_y^{-1} = \psi^\dagger \left( x, -y, z + \frac{1}{2} \right), \tag{5.95}$$

representing a layer with $C_z = -1$ at $z = 1/2$. Then, the Hamiltonian reads

$$\begin{aligned}
\hat{H} = \sum_{x,y,z \in \mathbb{Z}} \Bigg[ & \left( \psi^\dagger(x+1, y, z) \frac{\sigma_z + i\sigma_x}{2} \psi(x, y, z) + \psi^\dagger(x, y+1, z) \frac{\sigma_z + i\sigma_y}{2} \psi(x, y, z) + \text{h.c.} \right) \\
& + m\psi^\dagger(x, y, z) \sigma_z \psi(x, y, z) \\
& + \left( \psi^\dagger \left( x+1, -y, z + \frac{1}{2} \right) \frac{\sigma_z + i\sigma_x}{2} \psi \left( x, -y, z + \frac{1}{2} \right) \right. \\
& \left. + \psi^\dagger \left( x, -y-1, z + \frac{1}{2} \right) \frac{\sigma_z + i\sigma_y}{2} \psi \left( x, -y, z + \frac{1}{2} \right) + \text{h.c.} \right) \\
& + m\psi^\dagger \left( x, -y, z + \frac{1}{2} \right) \sigma_z \psi \left( x, -y, z + \frac{1}{2} \right) \Bigg] \\
= \sum_{x,y,z \in \mathbb{Z}} \Bigg[ & \left( \psi^\dagger(x+1, y, z) \frac{\sigma_z + i\sigma_x}{2} \psi(x, y, z) + \psi^\dagger(x, y+1, z) \frac{\sigma_z + i\sigma_y}{2} \psi(x, y, z) + \text{h.c.} \right) \\
& + \left( \psi^\dagger \left( x+1, y, z + \frac{1}{2} \right) \frac{\sigma_z + i\sigma_x}{2} \psi \left( x, y, z + \frac{1}{2} \right) \right. \\
& \left. + \psi^\dagger \left( x, y+1, z + \frac{1}{2} \right) \frac{\sigma_z - i\sigma_y}{2} \psi \left( x, y, z + \frac{1}{2} \right) + \text{h.c.} \right) \\
& + m\psi^\dagger(x, y, z) \sigma_z \psi(x, y, z) + m\psi^\dagger \left( x, y, z + \frac{1}{2} \right) \sigma_z \psi \left( x, y, z + \frac{1}{2} \right) \Bigg].
\end{aligned}$$

Let us introduce the $k$-space basis

$$\Psi^\dagger(k_x, k_y, k_z) \equiv \sum_{x,y,z \in \mathbb{Z}} \left( \psi^\dagger(x, y, z), \psi^\dagger \left( x, y, z + \frac{1}{2} \right) \right) e^{i(k_x x + k_y y + k_z z)}. \tag{5.96}$$

The glide and inversion operations are represented as

$$\begin{aligned}
\hat{G}_y \Psi^\dagger(k_x, k_y, k_z) \hat{G}_y^{-1} &= \sum_{x,y,z \in \mathbb{Z}} \left( \psi^\dagger \left( x, -y, z + \frac{1}{2} \right), \psi^\dagger(x, -y, z+1) \right) e^{i(k_x x + k_y y + k_z z)} \\
&= \Psi^\dagger(k_x, -k_y, k_z) \begin{pmatrix} 0 & e^{-ik_z} \\ 1 & 0 \end{pmatrix}_\tau, \\
\hat{I} \Psi^\dagger(k_x, k_y, k_z) \hat{I}^{-1} &= \sum_{x,y,z \in \mathbb{Z}} \left( \psi^\dagger(-x, -y, -z), \psi^\dagger \left( -x, -y, -z - \frac{1}{2} \right) \right) \sigma_z e^{i(k_x x + k_y y + k_z z)} \\
&= \Psi^\dagger(-k_x, -k_y, -k_z) \sigma_z \begin{pmatrix} 1 & 0 \\ 0 & e^{-ik_z} \end{pmatrix}_\tau.
\end{aligned}$$

Therefore, a representative Hamiltonian for the glide-symmetric $Z_2$ magnetic TCI phase with additional inversion symmetry is obtained as

$$H_{\tilde{\nu}}^{\mathrm{LC}}(\mathbf{k}) = (m + \cos k_x + \cos k_y)\sigma_z + \sin k_x \sigma_x + \sin k_y \sigma_y \tau_z. \tag{5.97}$$

Let us compute the indicator for $-2 < m < 0$. At the high-symmetry points, the Hamiltonians and the symmetry operators within the occupied states are shown in the following table:

| $P$ | $H(P)$ | $G_y(P)$ | $C_2(P)\vert_{\mathrm{occ}}$ |
|---|---|---|---|
| $\Gamma$ | $\sigma_z$ | $\tau_x$ | $-\tau_x$ |
| $Y$ | $-\sigma_z$ | $\tau_x$ | $\tau_x$ |
| $Z$ | $-\sigma_z$ | $\tau_x$ | $\tau_x$ |
| $C$ | $-\sigma_z$ | $\tau_x$ | $\tau_x$ |

.

Here an overall positive coefficients are omitted for simplicity. Therefore, the glide-$Z_2$ invariant becomes

$$(-1)^{\tilde{\nu}} = \prod_{i \in \mathrm{occ}} \frac{\zeta_i^-(\Gamma)\zeta_i^+(C)}{\zeta_i^-(Y)\zeta_i^+(Z)} = \frac{(-1) \times 1}{1 \times 1} = -1,$$

i.e., $\tilde{\nu} = 1$. As a result, $\tilde{\nu}$ is equal to 1 for $|m| < 2$, and $\tilde{\nu} = 0$ for other values of $m$.

#### 1.1.2: The Layer Construction (010; $\frac{1}{2}$) ($\tilde{\nu} = 0, C_y = 1$)

We here calculate the topological invariants for the LC (010; $\frac{1}{2}$). The model Hamiltonian is constructed by putting a Chern insulator with $C_y = 1$ on the $xz$ plane at $y = \frac{1}{2}$ given by Eq. (2.118) with inversion symmetry

$$\hat{I}\psi^\dagger(x, z)\hat{I}^{-1} = \psi^\dagger(-x, -z)\sigma_z.$$

We first take the length of the unit cell along the $z$ direction to be $1/2$, in order to make the model compatible with the glide symmetry. We then make copies by the translation $T_y$,

$$\hat{T}_y\psi^\dagger(x, y, z)\hat{T}_y^{-1} = \psi^\dagger(x, y + 1, z),$$

and the glide operation is given by

$$\hat{G}_y\psi^\dagger(x, y, z)\hat{G}_y^{-1} = \psi^\dagger\left(x, -y, z + \frac{1}{2}\right).$$

In momentum space, the Hamiltonian and symmetry operators are represented by

$$H(k_x, k_y, k_z) = \sin\frac{k_z}{2}\sigma_x + \sin k_x\sigma_y + \left(m + \cos k_x + \cos\frac{k_z}{2}\right)\sigma_z,$$
$$G_y(k_x, k_y, k_z) = e^{-ik_y}e^{-ik_z/2}, \qquad I(k_x, k_y, k_z) = e^{-ik_y}\sigma_z.$$

This is a two-band model, when the primitive vector along the $z$ axis to be $(0, 0, 1/2)$. One can formally regard the primitive vector along the $z$ axis to be $(0, 0, 1)$, which renders this model to a four-band model. Let us introduce the $k$-space basis

$$\Psi^\dagger(k_x, k_y, k_z) \equiv \sum_{x, y+\frac{1}{2}, z\in\mathbb{Z}} \left(\psi^\dagger(x, y, z), \psi^\dagger\left(x, y, z + \frac{1}{2}\right)\right) e^{i(k_x x + k_y y + k_z z)}. \tag{5.98}$$

Then the Hamiltonian is given by

$$H(k_x, k_y, k_z) = \left[\sin k_x\sigma_y + (m + \cos k_x)\sigma_z\right] + \left(\cos\frac{k_z}{2}\sigma_z + \sin\frac{k_z}{2}\sigma_x\right)\left(\cos\frac{k_z}{2}\tau_x + \sin\frac{k_z}{2}\tau_y\right),$$
$$G_y(k_x, k_y, k_z) = e^{-ik_y}\begin{pmatrix} 0 & e^{-ik_z} \\ 1 & 0 \end{pmatrix}_\tau, \qquad I(k_x, k_y, k_z) = e^{-ik_y}\sigma_z\begin{pmatrix} 1 & 0 \\ 0 & e^{-ik_z} \end{pmatrix}_\tau.$$

The $C_2$ rotation is

$$C_2(k_x, k_y, k_z) \equiv G_y(-k_x, -k_y, -k_z)I(k_x, k_y, k_z) = \sigma_z\begin{pmatrix} 0 & 1 \\ 1 & 0 \end{pmatrix}_\tau.$$

Let us compute the indicator for $-2 < m < 0$. At the high-symmetry points, the Hamiltonians and the symmetry operators within the occupied states are

| $P$ | $H(P)$ | $G_y(P)$ | $C_2(P)$ |
|---|---|---|---|
| $\Gamma$ | $(m+1)\sigma_z + \tau_x\sigma_z$ | $\tau_x$ | $\sigma_z\tau_x$ |
| $Y$ | $(m-1)\sigma_z + \tau_x\sigma_z$ | $\tau_x$ | $\sigma_z\tau_x$ |
| $Z$ | $(m+1)\sigma_z + \tau_x\sigma_z$ | $-\tau_x$ | $\sigma_z\tau_x$ |
| $C$ | $(m-1)\sigma_z + \tau_x\sigma_z$ | $-\tau_x$ | $\sigma_z\tau_x$ |

Therefore, the glide-$Z_2$ invariant is calculated as

$$(-1)^{\tilde{\nu}} = \prod_{i\in\mathrm{occ}} \frac{\zeta_i^-(\Gamma)\zeta_i^+(C)}{\zeta_i^-(Y)\zeta_i^+(Z)} = +1$$

for all the values of $m$ and it turned out to be trivial. On the other hand, this model has $C_y = +1$ for $-2 < m < 0$ and $C_y = -1$ for $0 < m < 2$ by construction.

## *1.2: Derivation of Tight-Binding Models for 15*

### 1.2.1: The Layer Construction (001; 0) ($\tilde{\nu} = 1, C_y = 0$)

The model Hamiltonian is constructed by putting a Chern insulator with $C_z = 1$ on the $xy$ plane at $z = 0$ with the inversion symmetry

$$\hat{I}\psi^\dagger(x, y)\hat{I}^{-1} = \psi^\dagger(-x, -y)\sigma_z,$$

given by Eq. (2.118), and by making copies by the glide transformation

$$\hat{G}_y\psi^\dagger(x, y, z)\hat{G}_y^{-1} = \psi^\dagger\left(x, -y, z + \frac{1}{2}\right). \tag{5.99}$$

The Hamiltonian reads

$$\begin{aligned}
\hat{H} = \sum_{x,y\in\mathbb{Z}/2,z\in\mathbb{Z}} \Bigg[ & \left(\psi^\dagger\left(x + \frac{1}{2}, y + \frac{1}{2}, z\right)\frac{\sigma_z + i\sigma_x}{2}\psi(x, y, z)\right. \\
& \left. + \psi^\dagger\left(x - \frac{1}{2}, y + \frac{1}{2}, z\right)\frac{\sigma_z + i\sigma_y}{2}\psi(x, y, z) + \text{h.c.}\right) \\
& + m\psi^\dagger(x, y, z)\sigma_z\psi(x, y, z) \\
& + \left(\psi^\dagger\left(x + \frac{1}{2}, -y - \frac{1}{2}, z + \frac{1}{2}\right)\frac{\sigma_z + i\sigma_x}{2}\psi\left(x, -y, z + \frac{1}{2}\right)\right. \\
& \left. + \psi^\dagger\left(x - \frac{1}{2}, -y - \frac{1}{2}, z + \frac{1}{2}\right)\frac{\sigma_z + i\sigma_y}{2}\psi\left(x, -y, z + \frac{1}{2}\right) + \text{h.c.}\right) \\
& + m\psi^\dagger\left(x, -y, z + \frac{1}{2}\right)\sigma_z\psi\left(x, -y, z + \frac{1}{2}\right)\Bigg] \\
= \sum_{x,y\in\mathbb{Z}/2,z\in\mathbb{Z}} \Bigg[ & \left(\psi^\dagger\left(x + \frac{1}{2}, y + \frac{1}{2}, z\right)\frac{\sigma_z + i\sigma_x}{2}\psi(x, y, z)\right. \\
& \left. + \psi^\dagger\left(x - \frac{1}{2}, y + \frac{1}{2}, z\right)\frac{\sigma_z + i\sigma_y}{2}\psi(x, y, z) + \text{h.c.}\right) \\
& + \left(\psi^\dagger\left(x + \frac{1}{2}, y - \frac{1}{2}, z + \frac{1}{2}\right)\frac{\sigma_z + i\sigma_x}{2}\psi\left(x, y, z + \frac{1}{2}\right)\right. \\
& \left. + \psi^\dagger\left(x - \frac{1}{2}, y - \frac{1}{2}, z + \frac{1}{2}\right)\frac{\sigma_z + i\sigma_y}{2}\psi\left(x, y, z + \frac{1}{2}\right) + h.c.\right) \\
& + m\psi^\dagger(x, y, z)\sigma_z\psi(x, y, z) + m\psi^\dagger\left(x, y, z + \frac{1}{2}\right)\sigma_z\psi\left(x, y, z + \frac{1}{2}\right)\Bigg].
\end{aligned}$$

Let us introduce the $k$-space basis

$$\Psi^\dagger(k_x,k_y,k_z) \equiv \sum_{x,y\in\mathbb{Z}/2, z\in\mathbb{Z}} \left(\psi^\dagger(x,y,z), \psi^\dagger\left(x,y,z+\frac{1}{2}\right)\right) e^{i(k_x x+k_y y+k_z z)}. \tag{5.100}$$

The glide and inversion operations are represented as

$$\begin{aligned}
\hat{G}_y\Psi^\dagger(k_x,k_y,k_z)\hat{G}_y^{-1} &= \sum_{x,y\in\mathbb{Z}/2, z\in\mathbb{Z}} \left(\psi^\dagger\left(x,-y,z+\frac{1}{2}\right), \psi^\dagger\left(x,-y,z+\frac{1}{2}\right)\right) e^{i(k_x x+k_y y+k_z z)} \\
&= \Psi^\dagger(k_x,-k_y,k_z)\begin{pmatrix} 0 & e^{-ik_z} \\ 1 & 0 \end{pmatrix}_\tau, \\
\hat{I}\Psi^\dagger(k_x,k_y,k_z)\hat{I}^{-1} &= \sum_{x,y\in\mathbb{Z}/2, z\in\mathbb{Z}} \left(\psi^\dagger(-x,-y,-z), \psi^\dagger\left(-x,-y,-z-\frac{1}{2}\right)\right) e^{i(k_x x+k_y y+k_z z)}\sigma_z \\
&= \Psi^\dagger(-k_x,-k_y,-k_z)\sigma_z\begin{pmatrix} 1 & 0 \\ 0 & e^{-ik_z} \end{pmatrix}_\tau.
\end{aligned}$$

In sum, the momentum space Hamiltonian and symmetry operators are

$$\begin{aligned}
H(k_x,k_y,k_z) &= \left(m+\cos\frac{k_x+k_y}{2}+\cos\frac{-k_x+k_y}{2}\right)\sigma_z \\
&\quad +\sin\frac{k_x+k_y}{2}\begin{pmatrix} \sigma_x & 0 \\ 0 & \sigma_y \end{pmatrix}+\sin\frac{k_x-k_y}{2}\begin{pmatrix} \sigma_y & 0 \\ 0 & \sigma_x \end{pmatrix} \\
&= \left(m+2\cos\frac{k_x}{2}\cos\frac{k_y}{2}\right)\sigma_z+\sin\frac{k_x}{2}\cos\frac{k_y}{2}(\sigma_x+\sigma_y)+\cos\frac{k_x}{2}\sin\frac{k_y}{2}(\sigma_x-\sigma_y)\tau_z, \\
G_y(k_x,k_y,k_z) &= \begin{pmatrix} 0 & e^{-ik_z} \\ 1 & 0 \end{pmatrix}_\tau, \quad I(k_x,k_y,k_z) = \sigma_z\begin{pmatrix} 1 & 0 \\ 0 & e^{-ik_z} \end{pmatrix}_\tau.
\end{aligned}$$

Let us compute the indicator for $-2 < m < 0$. The $C_2$ operation is

$$C_2(k_x,k_y,k_z) \equiv G_y(-k_x,-k_y,-k_z)I(k_x,k_y,k_z) = \sigma_z\tau_x. \tag{5.101}$$

At the high-symmetry points, the Hamiltonians and symmetry operators within the occupation states are given by the following:

| $P$ | $H(P)$ | $C_2(P)\vert_{\mathrm{occ}}$ | $G_y(P)\vert_{\mathrm{occ}}$ | $I(P)\vert_{\mathrm{occ}}$ |
|---|---|---|---|---|
| $\Gamma$ | $(m+2)\sigma_z$ | $\sigma_z\tau_x$ | $\tau_x$ | $\sigma_z$ |
| $Y$ | $(m-2)\sigma_z$ | $\sigma_z\tau_x$ | $\tau_x$ | $\sigma_z$ |
| $V$ | $m\sigma_z$ | $\sigma_z\tau_x$ | $\tau_x$ | $\sigma_z$ |

Therefore, the glide-$Z_2$ invariant is calculated as

$$(-1)^{\tilde{\nu}} = \prod_{i\in\mathrm{occ}} \zeta_i^+(\Gamma)\xi_i(V)\frac{\xi_i(V)}{\zeta_i^+(Y)} = -1,$$

i.e., $\tilde{\nu} = 1$. As a result, $\tilde{\nu}$ is equal to 1 to $|m| < 2$, and $\tilde{\nu} = 0$ for other values of $m$.

### 1.2.2: The Layer Construction $(\bar{1}10; 0)$ $(\tilde{\nu} = 0, C_y = 2)$

We here calculate the topological invariants for the LC $(\bar{1}10; 0)$. This layer should coexist with the layer $(110; 0)$. To construct the model Hamiltonian, we put a Chern insulator in Eq. (2.118) with $C = 1$ whose normal vector is along the $(\bar{1}10)$ direction, so that it includes the origin for inversion symmetry

$$\hat{I}\psi^\dagger(x, y, z)\hat{I}^{-1} = \psi^\dagger(-x, -y, -z)\sigma_z, \tag{5.102}$$

and we make copies by the glide transformation

$$\hat{G}_y\psi^\dagger(x, y, z)\hat{G}_y^{-1} = \psi\left(x, -y, z + \frac{1}{2}\right). \tag{5.103}$$

The Hamiltonian reads

$$\begin{aligned}
\hat{H} = \sum_{x,y\in\mathbb{Z}/2,z\in\mathbb{Z}} \Bigg[ & \left(\psi^\dagger(x, y, z+1)\frac{\sigma_z + i\sigma_x}{2}\psi(x, y, z) + \psi^\dagger\left(x + \frac{1}{2}, y + \frac{1}{2}, z\right)\frac{\sigma_z + i\sigma_y}{2}\psi(x, y, z) + \text{h.c.}\right) \\
& + m\psi^\dagger(x, y, z)\sigma_z\psi(x, y, z) \\
& + \left(\psi^\dagger\left(x, -y, z + \frac{1}{2}\right)\frac{\sigma_z + i\sigma_x}{2}\psi\left(x, -y, z + \frac{1}{2}\right)\right. \\
& \left. + \psi^\dagger\left(x + \frac{1}{2}, -y - \frac{1}{2}, z + \frac{1}{2}\right)\frac{\sigma_z + i\sigma_y}{2}\psi\left(x, -y, z + \frac{1}{2}\right) + \text{h.c.}\right) \\
& + m\psi^\dagger\left(x, -y, z + \frac{1}{2}\right)\sigma_z\psi\left(x, -y, z + \frac{1}{2}\right)\Bigg] \\
= \sum_{x,y\in\mathbb{Z}/2,z\in\mathbb{Z}} \Bigg[ & \left(\psi^\dagger(x, y, z+1)\frac{\sigma_z + i\sigma_x}{2}\psi(x, y, z) + \psi^\dagger\left(x + \frac{1}{2}, y + \frac{1}{2}, z\right)\frac{\sigma_z + i\sigma_y}{2}\psi(x, y, z) + \text{h.c.}\right) \\
& + \left(\psi^\dagger\left(x, y, z + \frac{1}{2}\right)\frac{\sigma_z + i\sigma_x}{2}\psi\left(x, y, z + \frac{1}{2}\right)\right. \\
& \left. + \psi^\dagger\left(x + \frac{1}{2}, y - \frac{1}{2}, z + \frac{1}{2}\right)\frac{\sigma_z + i\sigma_y}{2}\psi\left(x, y, z + \frac{1}{2}\right) + \text{h.c.}\right) \\
& + m\psi^\dagger(x, y, z)\sigma_z\psi(x, y, z) + m\psi^\dagger\left(x, y, z + \frac{1}{2}\right)\sigma_z\psi\left(x, y, z + \frac{1}{2}\right)\Bigg].
\end{aligned}$$

Let us introduce the $k$-space basis

$$\Psi^\dagger(k_x, k_y, k_z) \equiv \sum_{x,y\in\mathbb{Z}/2,z\in\mathbb{Z}} \left(\psi^\dagger(x, y, z), \psi^\dagger\left(x, y, z + \frac{1}{2}\right)\right) e^{i(k_x x + k_y y + k_z z)}. \tag{5.104}$$

The momentum space Hamiltonian and symmetry operators are represented

$$H(\mathbf{k}) = \begin{pmatrix} H^{(+)}\mathbf{k}) & \\ & H^{(-)}(\mathbf{k}) \end{pmatrix}_\tau,$$
$$H^{(\pm)}(\mathbf{k}) = \left(m + \cos k_z + \cos \frac{k_x \pm k_y}{2}\right)\sigma_z + \sin k_z \sigma_x \pm \sin \frac{k_x \pm k_y}{2}\sigma_y,$$
$$G_y(k_x, k_y, k_z) = \begin{pmatrix} 0 & e^{-ik_z} \\ 1 & 0 \end{pmatrix}_\tau, \quad I(k_x, k_y, k_z) = \sigma_z \begin{pmatrix} 1 & 0 \\ 0 & e^{-ik_z} \end{pmatrix}_\tau.$$

The $C_2$ operation is

$$C_2(k_x, k_y, k_z) \equiv G_y(-k_x, -k_y, -k_z)I(k_x, k_y, k_z) = \sigma_z \tau_x. \tag{5.105}$$

Let us compute the indicator for $-2 < m < 0$. At the high-symmetry points, the Hamiltonian and symmetry operators within the occupied states are given by the following:

| $P$ | $H(P)$ | $G_y\vert_{\text{occ}}$ | $I\vert_{\text{occ}}$ |
|---|---|---|---|
| $\Gamma$ | $(m+2)\sigma_z$ | $\tau_x$ | $\sigma_z$ |
| $Y$ | $m\sigma_z$ | $\tau_x$ | $\sigma_z$ |
| $V(V')$ | $(m+1)\sigma_z \mp \sigma_z\tau_z$ | | $\sigma_z$ |
| $M$ | $(m-2)\sigma_z$ | $-i\tau_y$ | $\sigma_z\tau_z$ |
| $L(L')$ | $(m-1)\sigma_z \mp \sigma_z\tau_z$ | | $\sigma_z\tau_z$ |

Therefore, the glide-$Z_2$ invariant is calculated as

$$(-1)^{\tilde{\nu}} = \prod_{i\in\text{occ}} \zeta_i^+(\Gamma)\xi_i(V)\frac{\xi_i(V)}{\zeta_i^+(Y)} = +1$$

for all the values of $m$ and it turned out to be trivial. On the other hand, this model has $C_y/2 = +1$ for $-2 < m < 0$ and $C_y/2 = -1$ for $0 < m < 2$ by construction.

## References


1. Song Z, Zhang T, Fang Z, Fang C (2018) Nat Commun 9(1):3530
2. Aroyo MI, Perez-Mato JM, Capillas C, Kroumova E, Ivantchev S, Madariaga G, Kirov A, Wondratschek H (2006) Z Krist-Cryst Mater 221(1):15
3. Hahn T (2002) International tables for crystallography, volume A: space group symmetry, 5th edn. Kluwer Academic Publishers, Dordrecht
4. Ono S, Watanabe H (2018) Phys Rev B 98:115150. https://doi.org/10.1103/PhysRevB.98.115150
5. Vanderbilt D (2018) Berry phases in electronic structure theory: electric polarization, orbital magnetization and topological insulators. Cambridge University Press, Cambridge

# Chapter 6
# Glide-Symmetric $Z_2$ Topological Crystalline Insulators in Magnetic Photonic Crystals

Manipulations for magnetic topological materials are an intriguing and promising topic in condensed matter physics. In particular, a glide-symmetric topological crystalline insulator (TCI) exhibits a $Z_2$ topological phase in class A hosting a single surface Dirac cone [1–3]. One of our targets for material realizations is bosonic systems. In optics, a new type of topological band-crossing points beyond the Weyl and Dirac points emerges. For example, at a generalized Dirac point [4], the bands are four-fold degenerate and they split into three or four bands along any direction, in contrast with a Dirac points, which generally splits into two sets of double degenerate bands along any direction. When the system is perturbed, this generalized Dirac point becomes line nodes, Weyl points, or opens a band gap [4, 5]. As an example, in the photonic crystal having a structure in the first blue phase of liquid crystals (BPI), by introducing magnetization to break time-reversal symmetry (TRS) the generalized Dirac point opens a band gap and the system goes into topological phases protected by glide symmetry [4]. On the other hand, in the double gyroid photonic crystal belonging to the same space group ***230*** (henceforth, we call a space group by its number in bold italic following in Ref. [6]) with the BPI photonic crystal, the system has Weyl points in the absence of TRS [5]. Nonetheless, the key to realize the topological phase has not been addressed.

In the present chapter, we show how this topological phase can be manipulated based on the relationship between space group representations and band structures. In the BPI photonic crystal, the gap opening at non-equivalent $P$ and $P'$ points in the Brillouin zone (BZ) makes the system gapped and topologically nontrivial. However, these $P$ and $P'$ points are not a time-reversal invariant momentum (TRIM), i.e., the band inversions at these points do not contribute to the change of the glide-$Z_2$ invariant that we have derived in Chap. 4 when inversion and glide symmetries coexist. To make clear this discrepancy, first we show the representations for the photonic band structure of the BPI photonic crystal to calculate the glide-$Z_2$ invariant. In this case, we have to pay attention to the $\Gamma$ point since it is a singularity in three-dimensional (3D) photonic crystals. As a consequence, we find that the key to determine whether the photonic crystal is topologically trivial or nontrivial relies on the $H$ point in the BZ. Then, we propose a way to design such a topological photonic



H. Kim, *Glide-Symmetric $Z_2$ Magnetic Topological Crystalline Insulators*,
Springer Theses, https://doi.org/10.1007/978-981-16-9077-8_6

crystal by introducing Wyckoff positions. Among eight Wyckoff positions in ***230***, we show that only one Wyckoff position is related to the photonic band structure realizing the glide-$Z_2$ magnetic TCI.

## 6.1 Previous Work: $Z_2$ Topological Photonic Crystal with Glide Symmetry in Breaking Time-Reversal Symmetry

We consider a 3D photonic crystal composed of four identical dielectric rods, constituting a body-centered cubic (BCC) lattice. A schematic illustration of the system is shown in Fig. 6.1a. The space group for this crystal turns out to be ***230*** ($Ia\bar{3}d$), which is nonsymmorphic, containing glide reflections and inversion.

When we set the dielectric constant of the rods to be $\varepsilon = 11$ and the radius of the rods to be $r = 0.13a$ where $a$ is the length of the cubic cell, the generalized Dirac point appears at two non-equivalent $P$ points in the BZ [4]. If the magnetization is present in these rods, the band structure has a band gap between bands 2 and 3 and it turns into the topological phase protected by glide symmetry [4]. To this end, in the dielectric tensor of the rod $\varepsilon$, we add off-diagonal imaginary terms $\kappa$

$$\varepsilon = \begin{pmatrix} \varepsilon_{\parallel} & \kappa & 0 \\ -\kappa & \varepsilon_{\parallel} & 0 \\ 0 & 0 & \varepsilon_{zz} \end{pmatrix}, \tag{6.1}$$

where $\varepsilon_{\parallel}^2 - |\kappa|^2 = \varepsilon_{zz}^2$ [4, 5], $\kappa$ is a non-zero imaginary number, and the $\hat{\mathbf{z}}$ axis here is taken along each rod. These off-diagonal imaginary terms are determined

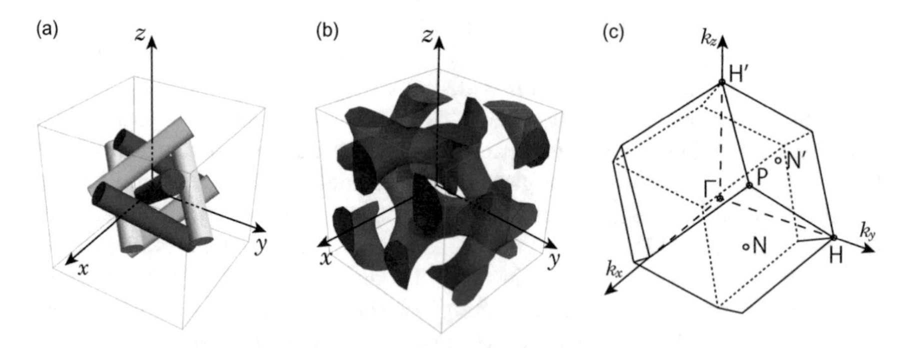


**Fig. 6.1** The BPI and double gyroid photonic crystals in the body-centered cubic (BCC) lattice and the Brillouin zone (BZ) of the BCC lattice. **a** The BPI photonic crystal. The cubic unit cell of length $a$ consists of four identical dielectric rods oriented along the BCC lattice vectors of (111) (red), (11$\bar{1}$) (yellow), (1$\bar{1}$1) (blue), and ($\bar{1}$11) (green), and they go through (0,0,0)$a$, (0,0.5,0)$a$, (0.5,0,0)$a$, and (0,0,0.5)$a$, respectively. **b** The double gyroid photonic crystal whose surfaces given by isosurfaces of $g(\mathbf{r})$ (blue) and $g(-\mathbf{r})$ (red) of Eq. (6.3) where dielectric region is $g(\mathbf{r}), g(-\mathbf{r}) > \lambda_{\text{iso}} = 1.1$. **c** The BZ of the BCC lattice and the high-symmetry points

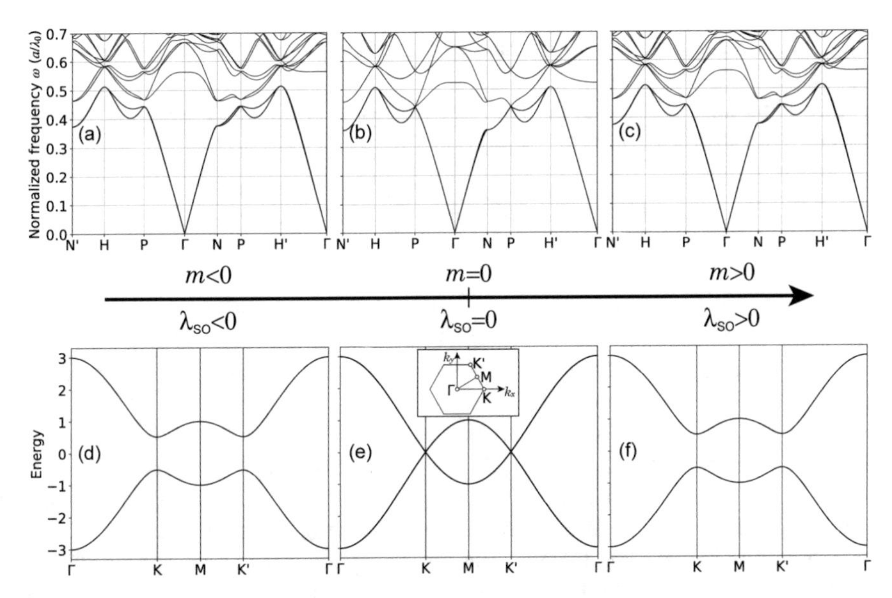


**Fig. 6.2** Band structures for **a–c** the BPI photonic crystal and **d–f** the Kane-Mele model with inversion symmetry. The band structures with magnetization where **a** $\varepsilon_{zz} = 11$ and $\kappa = 10i, 10i, -10i, -10i$ (magnetization $m$ along $-\hat{\mathbf{z}}$) and **c** with $\varepsilon_{zz} = 11$ and $\kappa = -10i, -10i, 10i, 10i$ (magnetization $m$ along $+\hat{\mathbf{z}}$) for the red, yellow, green and blue rods shown in Fig. 6.1a, respectively, are exactly the same. This situation is similar to the Kane-Mele model with **d** negative spin-orbit coupling $\lambda_{\text{so}}$ and **f** positive $\lambda_{\text{so}}$. The high-symmetry points for the BPI photonic crystal are shown in Fig. 6.1c, whereas those for the Kane-Mele model are shown in the inset of **e**, denoting the Brillouin zone of the honeycomb lattice

by the gyroelectric response of materials [7]. Similary, gyromagnetic response in ferrites gives off-diagonal terms of the permeability tensor [8]. Here we apply the magnetic field along the $\hat{\mathbf{z}}$ axis which is perpendicular to a glide plane. Generally, the in-plane diagonal components $\varepsilon_{\parallel}$ differ from the axial diagonal component $\varepsilon_{zz}$, and the off-diagonal component appears since the isotropy in the $xy$ plane is not broken. In Fig. 6.2a, c, we show the band structures of the BPI photonic crystal with $\kappa = \pm 10i$, $\varepsilon_{zz} = 11$. This value of $\kappa$ does not reflect real materials, but is not for demonstration of the behaviors of the gap. In real materials, this off-diagonal term $\kappa$ is small in the visible light, and can be larger in the radio frequency.

Compared with Fig. 6.2b for $\kappa = 0$, the band gap opens in (a) and (c) with $\kappa = \pm 10i$. The key to open the gap and to realize the topological phase is the splitting of the four-fold degenerate generalized Dirac point at the $P$ point. However, a role of the band inversion at the $P$ point in realizing the glide-symmetric $Z_2$ magnetic topological photonic crystal is not clear yet. If one supposes that the band inversion at the $P$ point contributes to the realization of the topological phase, the topological invariant should be different when the sign of the magnetization ($\kappa$) becomes opposite, and in addition, the surface states should appear at the projection of the

$P$ points. Nonetheless, these expectations contradict the behaviors in this photonic crystal.

This situation is similar to the Kane-Mele model (Fig. 6.2d–f), which is a model for graphene with spin-orbit interaction [9, 10]:

$$H = t \sum_{\langle ij \rangle} c_i^\dagger c_j + i\lambda_{\text{so}} \sum_{\langle\langle ij \rangle\rangle} c_i^\dagger (\mathbf{s} \cdot \hat{\mathbf{e}}_{ij}) c_j. \tag{6.2}$$

The first term is a nearest-neighbor hopping term, and the second term represents spin-orbit interaction for next-nearest neighbor hopping with $\hat{\mathbf{e}}_{ij} = (\mathbf{d}_{ij}^1 \times \mathbf{d}_{ij}^2)/|\mathbf{d}_{ij}^1 \times \mathbf{d}_{ij}^2|$ where $\mathbf{d}_{ij}^1$ and $\mathbf{d}_{ij}^2$ are vectors of the bonds constituting the next-nearest neighbor hopping from site $j$ to site $i$. When $\lambda_{\text{so}}$ is nonzero, Dirac points at the $K$ and $K'$ points in the BZ of graphene are gapped and the system becomes a 2D topological insulator (TI). This is consistent with the fact that the system becomes a 2D TI regardless of the sign of $\lambda_{\text{so}}$. In this model, as addressed in Ref. [11], the parities at four TRIMs in this model are given by $-1$ at one of three $M$ points and $+1$ at the other $M$ points and the $\Gamma$ point, and the $Z_2$ topological invariant for the 2D TI has a nontrivial value. Therefore, the wavefunctions at the $K$ points do not affect the $Z_2$ topological invariant for the 2D TI, and the band inversion for the topological phase is not at the $K$ points. Thus, in the presence of spin-orbit coupling, the system exhibits the 2D TI regardless of the sign of the spin-orbit coupling. The Dirac point for the edge states is pinned at the projection of $M$ point which is among the TRIMs.

Band inversion in the BPI photonic crystal we are interested in is similar to that in the Kane-Mele model. First, the magnetization is introduced to open the gap at the $P$ point, and regardless of the sign of the magnetization, the value of the topological invariant is the same. Furthermore, the surface states emerge at the place different from the projection of the $P$ point in bulk.

This is the reason why we revisit the topological photonic crystal ensured by glide symmetry. We will show that the opening of the band gap at the $P$ point does not contribute the glide-$Z_2$ invariant in magnetic systems. Instead, this photonic crystal realizes the topological phase which is independent of the wavefunction at the $P$ point, similar to the Kane-Mele model.

## 6.2 Symmetry Consideration for Band Theory of Glide-Symmetric $Z_2$ Magnetic Topological Photonic Crystal

In this section, we discuss what determines the topological phase in the photonic crystal from the symmetry viewpoint. We first investigate the irreducible representations (irreps) for the glide-symmetric $Z_2$ magnetic topological photonic crystal. Then, we calculate the glide-$Z_2$ invariant by using our new formula we derived in

Chap. 4. In arguing these points, we should pay attention to singularity at the Γ point which naturally arises in 3D photonic crystal systems.

### 6.2.1 Representations at the High-Symmetry Points

We start with deriving the set of the irreps of the eigenmodes at the high-symmetry points in the BZ. There are four kinds of high-symmetry points Γ, $H$, $N$, and $P$, in the BZ of the BCC lattice. Among the four kinds of high-symmetry points, the three kinds of points Γ, $H$, and $N$ are TRIMs in BZ, but $P$ is not. Even though $P$ is not a TRIM, the bands at the $P$ point should be degenerate in the presence of TRS. Suppose the combined symmetry $\mathcal{T}' = \mathcal{T} G_y$ where $\mathcal{T}$ is time-reversal operation. As $\mathcal{T}'$ is an antiunitary symmetry, $(\mathcal{T}')^2 = G_y^2 = T_z = -1$ at the $P$ point, and the band structure always appear with double degeneracy at the $P$ point.

Our aim is the following. We first start with a vacuum having a homogeneous dielectric constant $\varepsilon = 1$. Thereby, the eigenmodes behave as the electromagnetic plane waves with a linear polarization in free space. From symmetry considerations for these plane wave bases, we have reducible representations at high-symmetry points which characterize the eigenmodes at $\varepsilon = 1$ (Fig. 6.3a). Next, from the vacuum with $\varepsilon = 1$, we gradually the dielectric constant away from $\varepsilon = 1$ in the dielectrics to describe a photonic crystal. The reducible representations should be decomposed into irreps, and they split when the photonic crystal has a dielectric constant away from $\varepsilon = 1$. Therefore, we can finally get a set of irreps of the eigenmodes at the high-symmetry points in the BZ.

Let us consider the six points $\mathbf{k} = \pm b\mathbf{e}_i\,(i = x, y, z)$, $b = 2\pi/a$, and we call them $H^{(0)}$ points. $\mathbf{e}_x$, $\mathbf{e}_y$, and $\mathbf{e}_z$ denote the unit vectors along $x$, $y$, and $z$ directions, respectively. They are the lowest wavenumbers among the $H$ points. At these points the lowest band has the same frequency in free space, whose basis functions for a reducible representation are formed by the 12 plane wave bases. At first, we make clear the characters of the symmetry operations at the $H$ point by counting the number how many of $H^{(0)}$ points are invariant under a given point-group operation. For each of the $H^{(0)}$ points, it is unchanged under the identity transformation, a twofold ($C_2$) rotation, a fourfold ($C_4$) rotation, and a $IC_2'$ mirror reflection. All the other symmetry operations change all $\mathbf{k}$ on the $H^{(0)}$ points into $\mathbf{k} + \mathbf{G}$ with a nonzero reciprocal lattice vector $\mathbf{G}$. Because there are two independent polarization modes at each point, the character, i.e., the trace of the symmetry operations operating onto these two modes, is given by 2 for the identity transformation, $-2$ for the $C_2$ rotation, 0 for the $C_4$ rotation, and 0 for the $IC_2'$ mirror reflection. Therefore, we obtain the characters for the reducible representation at the $H^{(0)}$ point in Table 6.4 in Sect. 6.4.

Then we apply the compatibility relations. If these modes for the reducible representation are perturbed, such as changing the dielectric constant in the photonic crystal, they have different frequencies characterized by the irreps. By using the character table for the irreps of the $H$ point summarized in Table 6.6 (Sect. 6.4) and

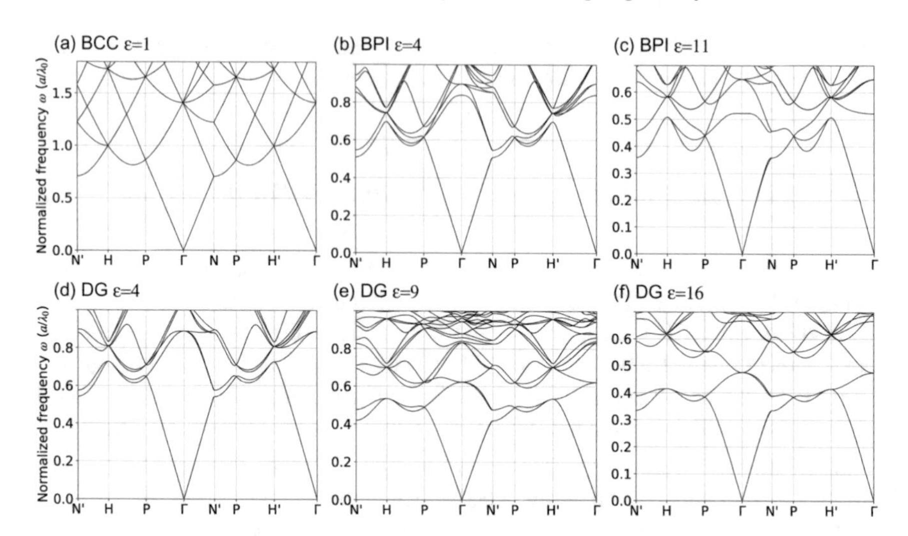


**Fig. 6.3** Evolutions of band structures for BPI and double gyroid (DG) photonic crystals. Band structures for **a** $\varepsilon = 1$, which is an air band structure embedded into the Brillouin zone of the body-centered cubic (BCC) photonic crystal. Band structures are shown for the BPI photonic crystal **b** at $\varepsilon = 4$, **c** at $\varepsilon = 11$, and for the DG photonic crystal **d** at $\varepsilon = 4$, **e** at $\varepsilon = 9$, **f** at $\varepsilon = 16$. There is **a** a twelve-fold, **b**, **c** two-fold, and **d–f** four-fold degeneracy in the lowest bands at the $H$ point

use the details of symmetry operations in Ref. [12], the eigenmodes with the lowest frequency at the $H$ points are split as $H_1 + H_2H_3 + H_4$ in the presence of TRS. This result is also confirmed by numerical calculations with changing the dielectric constant $\varepsilon$ in Fig. 6.3.

In a same manner, one can classify the irreps for the lowest and the higher frequencies at the high-symmetry points. These results are summarized in Table 6.4 in Sect. 6.4. Note that the lowest frequency $\omega = 0$ at the $\Gamma$ point has a singularity. As remarked in Ref. [13], eigenmodes at this singular point may not form representations of the symmetry of the system.

Let us check how the band structure in the BPI photonic crystal forms the irreps by performing numerical calculations for the band structures when the dielectric constant $\varepsilon$ in photonic crystal is evolved from 1 (Fig. 6.3a) via 4 (Fig. 6.3b) to 11 (Fig. 6.3c). At $\varepsilon = 1$, where the eigenmodes are the plane waves in free space, the degeneracies at the high-symmetry points completely agree with the above argument (Table 6.4). Then, those degenerate points are split into irreps in Table 6.4 when $\varepsilon$ is increased. In this case, the irreps of the lowest bands at high-symmetry points are given by $H_1$, $N_1$, and $P_3$ in Table 6.1. These irreps remain the lowest bands even up to $\varepsilon = 11$.

We also consider the double gyroid photonic crystal in Fig. 6.1b whose space group is ***230*** and is the same as the BPI photonic crystal. Nonetheless, this photonic crystal realizes a Weyl semimetal phase by magnetization and is not gapped [5]. A single gyroid surface is given by the isosurfaces of

**Table 6.1** A set of irreducible representations at high-symmetry points $H$, $N$, and $P$ for the lowest bands in BPI and double gyroid photonic crystals

| High-symmetry point | $H$ | $P$ | $N$ |
|---|---|---|---|
| BPI photonic crystal | $H_1$ | $P_3$ | $N_1$ |
| Double gyroid photonic crystal | $H_2 H_3$ | $P_1 P_2$ | $2N_1$ |

$$g(\mathbf{r}) = \sin(2\pi x/a)\cos(2\pi y/a) + \sin(2\pi y/a)\cos(2\pi z/a) + \sin(2\pi z/a)\cos(2\pi x/a), \quad (6.3)$$

where $a$ is the lattice constant in the BCC lattice [14], and the double gyroid surface is given by the combination of $g(\mathbf{r})$ and its counterpart by space inversion $g(-\mathbf{r})$.[1] We set the dielectric region as $g(\mathbf{r}), g(-\mathbf{r}) > \lambda_{\text{iso}}$ with $\lambda_{\text{iso}} = 1.1$ (Fig. 6.1b). We calculate its band structures when the dielectric constant $\varepsilon$ is changed from 1 (Fig. 6.3a) via 4 (Fig. 6.3d) and 9 (Fig. 6.3e) to 16 (Fig. 6.3f). The irreps of the four lowest bands at high-symmetry points are given by $H_2 H_3$, $2N_1$, and $P_1 P_2$ which are summarized in Table 6.1. This set of the irreps are different from the BPI photonic crystals.

### 6.2.2 Singularity of Photonic Bands

Before we calculate the glide-$Z_2$ invariant for this photonic crystal, we check whether the glide-$Z_2$ invariant can be defined in a 3D photonic crystal, because the formula of the glide-$Z_2$ invariant contains the integral of Berry curvature at $\Gamma$, even though $\Gamma$ has a singularity. Nonetheless, we find that the integral of Berry curvature including $\Gamma$ can be defined, i.e., we can define the glide-$Z_2$ invariant regardless of the singularity.

We can deal with this singularity as an adiabatic process, even though Maxwell equations in free space do not lead to electromagnetic waves at $\omega = |\mathbf{k}| = 0$ as mentioned in Ref. [13]. Around this singularity, two gapless modes behave in the same way as plane waves under symmetry operations. Thus, the two gapless photons have the same eigenvalues and eigenvectors as the plane waves of the vectorial electric field as remarked in Ref. [13].

Let us consider the glide operation $\hat{G}_y : (x, y, z) \to (x, -y, z + (c/2))$. Along the line $\Gamma$-$H$, on the glide-invariant plane $k_y = 0$, only one of the two gapless photons flips sign under the glide reflection of $\hat{G}_y$. Due to $(\hat{G}_y)^2 = e^{-ik_z c}$, the eigenvectors for the two gapless photons along the line $\Gamma$-$H$ on the glide-invariant plane $k_y = 0$ are $(E_x, 0, E_z)$ with the eigenvalue $+e^{-ik_z c/2}$ and $(0, E_y, 0)$ with $-e^{-ik_z c/2}$. The Berry curvature on the glide-invariant plane $k_y = 0$, expressed in terms of derivatives with respect to $k_x$ and $k_z$, is always zero, and the integral of the Berry curvature can be defined on this plane. Namely, the glide-$Z_2$ invariant is defined regardless of the existence of the singularity.

[1] In Fig. 6.1b, the surfaces of blue and red correspond to $g(\mathbf{r})$ and $g(-\mathbf{r})$, respectively.

### 6.2.3 The Glide-$Z_2$ Invariant for the Photonic Crystal

As ***15*** is one of the $t$-subgroups of ***230*** [6], here we directly calculate the glide-$Z_2$ invariant by using Eq. (4.81) instead of Eq. (4.82). The reason we use Eq. (4.81) is because the $\Gamma$ point may not form a representation, but we can calculate the $C_2$-eigenvalue at this point from the previous subsection. From correlations between the irreps of ***15*** and those of ***230***, the high-symmetry points $\Gamma$, $Y$, and $V$ in ***15*** emerging in Eq. (4.81) are projected from $\Gamma$, $H$, and $N$ in ***230***. Suppose there exists a band gap between the second and third bands when TRS is broken by applying a magnetic field. First, we consider the $C_2$ eigenvalue of the $g_+$ sector at the $\Gamma$ point. The eigenstate of the $g_+$ sector below the band gap at this point is given by $(E_x, 0, E_z)$, as we have shown in the previous subsection, and the eigenvalue of $C_2$ rotation along $y$ axis is immediately given by $-1$. Next, the two lowest bands form the irreps $N_1$ at the $N$ point and $H_1$ at the $H$ point in the BPI photonic crystal. The parities for those irreps are $\pm 1$ and the $C_2$ eigenvalue of $g_\pm$-sector is $-1$ as shown in Table 6.6. Therefore, the glide-$Z_2$ invariant given by Eq. (4.81) is

$$(-1)^\nu = \prod_i \zeta_i^+(\Gamma)\xi_i(V)\frac{\xi_i(Y)}{\zeta_i^+(Y)} = (-1)(-1)(+1)(-1)\frac{(+1)(-1)}{(-1)} = -1, \quad (6.4)$$

which means that the photonic crystal is topologically nontrivial.

One may wonder the role of the band inversions at the $P$ point, because this topological phase is realized by the band gap between the second and third bands at the $P$ points when the photonic crystal has magnetization. In fact, as shown in Eqs. (4.69) and (4.81), the $P$ point in ***230***, which corresponds to the $L$ point in ***15*** on the $k_y = \pi$ plane, does not contribute to the glide-$Z_2$ invariant. Therefore, regardless of the band inversions at the $P$ point, the irreps of the two lowest bands determine the nontrivial topological phases, similar to the Kane-Mele model of Eq. (6.2). Namely, when the band gap is present between the second and third bands, and the two lowest bands have the same set of the irreps as the present case, i.e., $N_1$ at $N$ and $H_1$ at $H$ in this case, the system always exhibits the $Z_2$ magnetic topological phase ensured by glide symmetry.

In fact, the irrep characterizing the lowest bands at the $H$ point plays a crucial role to determine whether the system is topologically trivial or nontrivial in this case, because the irreps for the lowest bands at the $N$ point and these at the $\Gamma$ point are fixed. To this end, assume that $H_1$ characterizes the lowest bands at the $H$ point. The compatibility relations via $\Lambda$ to the $P$ point, which is a threefold-rotation invariant line, yield the irrep $P_3$ for the lowest bands at the $P$ point because two 1D irreps $\Lambda_1$ and $\Lambda_2$, which is compatible with $H_1$, are only compatible with $P_3$ shown in Table 6.5 in Sect. 6.4. Therefore, the lowest bands with $H_1$ accompanies those with $P_3$ at the $P$ point. On the other hand, when the lowest bands at the $H$ point form $H_2H_3$, those at the $P$ point simultaneously form $P_1P_2$, due to the compatibility relations $H_2$, $H_3 \to \Lambda_3$ and $P_1$, $P_2 \to \Lambda_3$ (Table 6.5). Note that the former case with the lowest bands at $H$ being $H_1$ emerges in the BPI photonic crystal [4] (Fig. 6.3b, c), and the latter case with $H_2H_3$ emerges in the double gyroid photonic crystal [5] (Fig. 6.3d–f). When TRS is broken by applying a magnetic field, the double gyroid photonic crystal hosts Weyl points instead of opening the band gap. Therefore, the only chance for the photonic crystal with ***230*** to realize the glide-$Z_2$ TCI occurs when the lowest bands at the $H$ point is characterized by the irrep $H_1$. Then when the gap opens between the second and third bands, it becomes the glide-$Z_2$ TCI.

## 6.3 Manipulations for Topological Photonic Crystals with Glide Symmetry

In this section, we propose how topological photonic crystals are realized based on the representation theory and Wyckoff positions.

### *6.3.1 Perturbation Theory*

As we argued in the previous section, the irreps for the lowest bands at the $H$ point for the BPI photonic crystal and double gyroid photonic crystal are unchanged from a smaller value of the dielectric constant to a larger value. In order to see the level splitting at the $H$ point for a small value of $\varepsilon$, we can employ the perturbation theory in the dielectric function. When we add a small perturbation of the dielectric function $\delta\varepsilon(\mathbf{r})$, the frequency shift $\delta\omega$ is given as follows [15]:

$$\delta\omega = -\frac{\omega}{2}\frac{\int d^3\mathbf{r}\ \delta\varepsilon(\mathbf{r})|\mathbf{E}(\mathbf{r})|^2}{\int d^3\mathbf{r}\ \varepsilon(\mathbf{r})|\mathbf{E}(\mathbf{r})|^2}. \tag{6.5}$$

This relation implies that the more the electric fields concentrate in the region of $\delta\varepsilon(\mathbf{r})$, i.e., in the dielectrics, the more frequency shift occurs.

To this end, let us construct the plane wave eigenmodes from the irreps since we are interested in the case $\varepsilon(\mathbf{r}) \gtrsim 1$. Here we use the representation theorem

$$\psi_i = \sum_R \chi_i(R)\hat{R}\psi, \tag{6.6}$$

where $\chi_i(R)$ is the character for the symmetry operation of $R$, $\hat{R}$ is a symmetry operator, and $\psi$ is a basis function. There are 12 plane wave basis functions at the lowest frequency among the $H$ points. By using those basis functions, one can have a set of the eigenstates with the corresponding irreps $H_1$, $H_2$, $H_3$, and $H_4$, respectively, as summarized in Table 6.2. Then, we can calculate the frequency shift by using Eq. (6.5), if we know the spatial density of $\delta\varepsilon(\mathbf{r})$ corresponding to a photonic crystal. Nevertheless, one cannot determine the difference of the frequency shifts between

**Table 6.2** A set of eigenmodes with plane waves at the $H^{(0)}$ point based on the characters of irreducible representations in Table 6.6 by using Eq. (6.6). $\hat{\mathbf{x}}$, $\hat{\mathbf{y}}$, and $\hat{\mathbf{z}}$ denote the unit vectors along $x$, $y$, and $z$ directions, respectively

| | |
|---|---|
| $H_1$ | $\psi_{H_1}^1 = \cos\left(\frac{2\pi}{a}x\right)\hat{\mathbf{y}} + \cos\left(\frac{2\pi}{a}y\right)\hat{\mathbf{z}} + \cos\left(\frac{2\pi}{a}z\right)\hat{\mathbf{x}}$<br>$\psi_{H_1}^2 = \cos\left(\frac{2\pi}{a}x\right)\hat{\mathbf{z}} + \cos\left(\frac{2\pi}{a}y\right)\hat{\mathbf{x}} + \cos\left(\frac{2\pi}{a}z\right)\hat{\mathbf{y}}$ |
| $H_2$ | $\psi_{H_2}^1 = \cos\left(\frac{2\pi}{a}x\right)\hat{\mathbf{y}} + e^{2\pi i/3}\cos\left(\frac{2\pi}{a}y\right)\hat{\mathbf{z}} + e^{-2\pi i/3}\cos\left(\frac{2\pi}{a}z\right)\hat{\mathbf{x}}$<br>$\psi_{H_2}^2 = \cos\left(\frac{2\pi}{a}x\right)\hat{\mathbf{z}} + e^{2\pi i/3}\cos\left(\frac{2\pi}{a}y\right)\hat{\mathbf{x}} + e^{-2\pi i/3}\cos\left(\frac{2\pi}{a}z\right)\hat{\mathbf{y}}$ |
| $H_3$ | $\psi_{H_3}^1 = \cos\left(\frac{2\pi}{a}x\right)\hat{\mathbf{y}} + e^{-2\pi i/3}\cos\left(\frac{2\pi}{a}y\right)\hat{\mathbf{z}} + e^{2\pi i/3}\cos\left(\frac{2\pi}{a}z\right)\hat{\mathbf{x}}$<br>$\psi_{H_3}^2 = \cos\left(\frac{2\pi}{a}x\right)\hat{\mathbf{z}} + e^{-2\pi i/3}\cos\left(\frac{2\pi}{a}y\right)\hat{\mathbf{x}} + e^{2\pi i/3}\cos\left(\frac{2\pi}{a}z\right)\hat{\mathbf{y}}$ |
| $H_4$ | $\psi_{H_4}^1 = \sin\left(\frac{2\pi}{a}x\right)\hat{\mathbf{y}},\ \psi_{H_4}^2 = \sin\left(\frac{2\pi}{a}x\right)\hat{\mathbf{z}}$<br>$\psi_{H_4}^3 = \sin\left(\frac{2\pi}{a}y\right)\hat{\mathbf{z}},\ \psi_{H_4}^4 = \sin\left(\frac{2\pi}{a}y\right)\hat{\mathbf{x}}$<br>$\psi_{H_4}^5 = \sin\left(\frac{2\pi}{a}z\right)\hat{\mathbf{x}},\ \psi_{H_4}^6 = \sin\left(\frac{2\pi}{a}z\right)\hat{\mathbf{y}}$ |

the electric fields with $H_1$, $H_2$, and $H_3$ in Table 6.2, because the distribution of $|\mathbf{E}|^2$ in Eq. (6.5) is the same for $H_1$, $H_2$, and $H_3$. Namely, from the plane wave bases, because the wave functions does not localize in the region of nonzero $\delta\epsilon(\mathbf{r})$ in general, it is a challenging problem to capture the frequency shifts.

### *6.3.2 Photonic Band Structures Based on Wyckoff Positions*

In electronic systems, Wyckoff positions classify spatial locations consistent with a given space group and by putting orbitals with various symmetry properties on each site of the Wyckoff positions, one can exhaust all the band structures of various topological nature. Inspired from this, we adopt the concept of the Wyckoff position into our theoretical analysis.

There are 8 Wyckoff positions in ***230*** and their site symmetries are summarized in Table 6.3 [6]. Our objective is to answer which Wyckoff positions correspond to the band structure of the BPI or that of the double gyroid photonic crystals. To this end, we numerically calculate the band structure by putting dielectric spheres on a given Wyckoff position and enlarging the radius of the dielectric spheres. When the dielectric spheres are absent, the light propagates in the air whose band structure is the same in Fig. 6.3a, but if the dielectric spheres have a finite radius, we can observe the evolution of the band structure.

First, we put dielectric spheres on the Wyckoff position labeled $16a$. We here set the dielectric constant to be $\varepsilon = 12$. We also set the maximum radius of the dielectric sphere to be the radius when they touch each other. In this case, in enlarging the radius of the dielectric spheres and eventually touching each other, the band structure hardly changes from that for the air. Recall that Eq. (6.5) implies the degree of concentration of the electric fields in the dielectric regions. The fact that the frequency shifts between the wavefunctions with $H_1$, $H_2 H_3$, and $H_4$ in the case of the Wyckoff position $16a$

**Table 6.3** Wyckoff positions in ***230***, their site symmetries, and the irreducible representations (irreps) for the lowest bands with $\varepsilon = 12$ at the $H$ point, when dielectric spheres at each Wyckoff position are enlarged. Detailed coordinates of each Wyckoff position are summarized in Table 6.7

| Wyckoff positions | $16a$ | $16b$ | $24c$ | $24d$ | $32e$ | $48f$ | $48g$ |
|---|---|---|---|---|---|---|---|
| Site symmetry | $\bar{3}(C_{3i})$ | $32(D_3)$ | $222(D_2)$ | $\bar{4}(S_4)$ | $3(C_3)$ | $2(C_2)$ | $2(C_2)$ |
| Irreps for the lowest band at the $H$ point | $H_1H_2H_3H_4$ | $H_2H_3$ | $H_2H_3$ | $H_1H_2H_3$ | $H_1$ | $H_1H_2H_3$ | $H_2H_3$ |

are not conspicuous implies that they localize on the dielectric spheres at the Wyckoff position $16a$ to the same degree.

In a similar manner, we can calculate the corresponding band structures for dielectric spheres at other Wyckoff positions in ***230***. As a consequence, we find that the Wyckoff positions labeled by $16b$, $24c$, and $48g$ generate band structure similar to that from the double gyroid photonic crystal and the Wyckoff position labeled by $32e$ generates band structure similar to that from the BPI photonic crystal. Namely, the irreps for the lowest bands at the $H$ point are given by $H_2H_3$ for $16b$, $24c$, and $48g$, and by $H_1$ for $32e$. For the Wyckoff position $24d$, the lowest eight bands at the $N$ point and those at the $P$ point hardly split, and at the $H$ point, the eigenmodes for $H_1$ and $H_2H_3$ are also very close to each other that are away from those for $H_4$. The Wyckoff position $48f$ hardly changes the band structure from the air. We skip the most general sites $96h$ with the site symmetry $1(C_1)$. Remarkable cases are depicted in Fig. 6.4. Here we set the dielectric constant as $\varepsilon = 12$ and the radius of dielectric spheres as $r = \sqrt{3}/8$ for $16a$ (Fig. 6.4a) and $r = 0.18$ for $16b$ (Fig. 6.4b) with the lattice constant being unity.

Let us analyze these results. The site symmetry 3 for $32e$ is a subgroup of $\bar{3}$ for $16a$ and 32 for $16b$, and the site symmetry 2 for $48f$ and $48g$ is a subgroup of 32 for $16b$, 222 for $24c$, and $\bar{4}$ for $24d$. The Wyckoff position $32e$ lies on the four threefold ($C_3$) rotation axes along the (111), ($\bar{1}11$), ($1\bar{1}1$), and ($11\bar{1}$) directions connecting neighboring $16a$ and $16b$ sites. The BPI photonic crystal with $H_1$ being the lowest band indeed corresponds to the Wyckoff position $32e$. This is reasonable since from Table 6.3, we see that $32e$ is the best candidate to make $H_1$ to be the lowest band at the $H$ point. Here we recall that photonic crystals with $H_1$ being the lowest band become a glide-$Z_2$ TCI phase when the TRS is broken.

On the other hand, $48g$ exists between neighboring $16b$ and $24c$ sites, and it corresponds to the planes perpendicular to four $C_3$ rotation axes along the (111), ($\bar{1}11$), ($1\bar{1}1$), and ($11\bar{1}$) directions. Let us recall the double gyroid photonic crystal given by $g(\mathbf{r}), g(-\mathbf{r}) > \lambda_{\text{iso}}$ where $g(\mathbf{r})$ is defined in Eq. (6.3). The isosurface function $g(\mathbf{r})$ has a maximum value 1.5 when $\mathbf{r} = (x, y, z)$ is equal to the Wyckoff position $16b$. By changing $\lambda_{\text{iso}}$ smaller from $\lambda_{\text{iso}} = 1.5$, the dielectric region for the double gyroid photonic crystal broadens from the sites of $16b$ along $48g$, i.e., along the planes perpendicular to the four $C_3$ rotation axes. Note that the sites

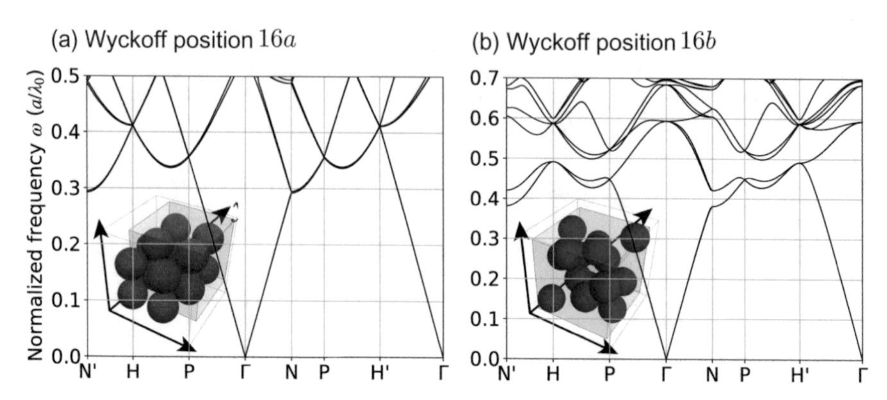


**Fig. 6.4** The photonic band structures at $\varepsilon = 12$ by putting dielectric spheres on the Wyckoff positions **a** 16*a* and **b** 16*b*. **a** Band structure with dielectric spheres with the radius $\sqrt{3}/8$ at the Wyckoff position 16*a*. It is hardly changed from an air band structure of a body-centered cubic photonic crystal. **b** Band structure with dielectric spheres with the radius 0.18 at the Wyckoff position 16*b*, not touching each other. The band structure is similar to that of the double gyroid photonic crystal. Here we set the lattice constant to be unity

of 16*a* and 24*d* always make $g(\mathbf{r})$ to be zero. Thus, it totally agrees with the fact that the band structures reproduce that of the double gyroid photonic crystal when 16*b*, 24*c* or 48*g* is dominant since 48*g* eventually reaches 24*c* from 16*b*. Moreover, 48*f* connects neighboring 24*c* and 24*d* sites where $C_2$ axes along $x$, $y$, $z$-directions lie. Therefore, we expect that the wavefunctions with $H_1$ might be localized along (111), $(\bar{1}11)$, $(1\bar{1}1)$, and $(11\bar{1})$ directions, which are the $C_3$ rotation axes, while those with $H_2H_3$ prefer the planes perpendicular to the $C_3$ rotation axes. As a consequence, we can expect the glide-$Z_2$ TCI phase in the case of the Wyckoff position 32*e* being dominant, because other cases never have the lowest bands with $H_1$ at the $H$ point.

### 6.3.3 Designing for Topological Photonic Crystals Based on Wyckoff Positions

We have shown that dielectric spheres located at the Wyckoff position 32*e* in ***230*** generates band structures with a topological gapped phase same as that in the BPI photonic crystal, while those at 16*b* in ***230*** generates those with Weyl nodes same as the double gyroid photonic crystal in the absence of TRS. Then, the question arises: is it possible to manipulate topological photonic crystals based on information of the Wyckoff position? To answer this question, we focus on Wyckoff positions with the least multiplicity, i.e., 16*a* and 16*b* because we can generate 32*e* by connecting those Wyckoff positions along (111), $(\bar{1}11)$, $(1\bar{1}1)$, and $(11\bar{1})$ directions.

First, we examine the photonic band structures by putting dielectric spheres on both sites of 16*a* and 16*b* and by changing their radii, with keeping the spheres

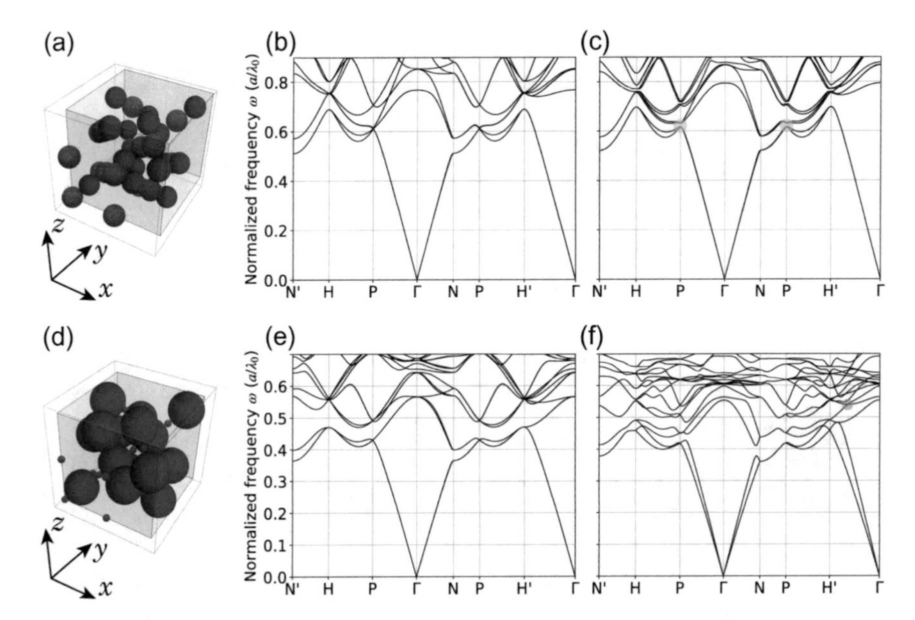


**Fig. 6.5** Configurations of photonic crystals with dielectric spheres at the Wyckoff positions $16a$ and $16b$, and the corresponding band structures with and without time-reversal symmetry (TRS). **a** We put dielectric spheres with the radius $\sqrt{3}/16$ on the Wyckoff positions $16a$ and $16b$. In this configuration, the band structures **b** with and **c** without TRS are qualitatively similar to those of the BPI photonic crystal and there is a gap between the second and third bands when the magnetization is present in **c**. **d** At the Wyckoff position $16b$, we put blue spheres with the radii $5\sqrt{3}/48$ and red spheres with the radii $\sqrt{3}/48$. The band structures **e** with and **f** without TRS are similar to those of the double gyroid photonic crystal and the Weyl node is the only touching point between the fourth and fifth bands in the absence of TRS in **f**

touching each other. Namely, when $r_a$ and $r_b$ denote the radii of dielectric spheres at the sites of $16a$ and $16b$, we change $r_a$ and $r_b$ under the constraint $r_a + r_b = \sqrt{3}/8$. Let us start with $r_a = r_b = \sqrt{3}/16$. The configuration is depicted in Fig. 6.5a where the gray cuboid represents a conventional unit cell $0 \leq x, y, z < 1$, and the band structures with and without TRS are shown in Fig. 6.5b, c, respectively. The band structure with TRS (Fig. 6.5b) qualitatively agrees with that of the BPI photonic crystal (Fig. 6.2b). Next we introduce magnetization while preserving glide and $C_3$ rotational symmetries, which is realized by putting $\varepsilon_{zz} = 12$, $\kappa = -11i$ for dielectric spheres along (111) and $(11\bar{1})$ directions, and $\kappa = 11i$ for those along $(\bar{1}11)$ and $(1\bar{1}1)$ on Eq. (6.1). It opens the gap at the $P$ point and the system goes into a gapped system. Since it has the same set of irreps as the BPI photonic crystal, this gapped system should be the glide-$Z_2$ TCI phase. If we make the system similar to the photonic crystal with spheres at $16b$, by putting $5r_a = r_b = 5\sqrt{3}/48$ depicted in Fig. 6.5d, the band structures with and without TRS are given in Fig. 6.5e, f, respectively. In this case, although the fourth and fifth bands are closer than those of the double gyroid photonic crystal, the band structure with TRS basically agrees with

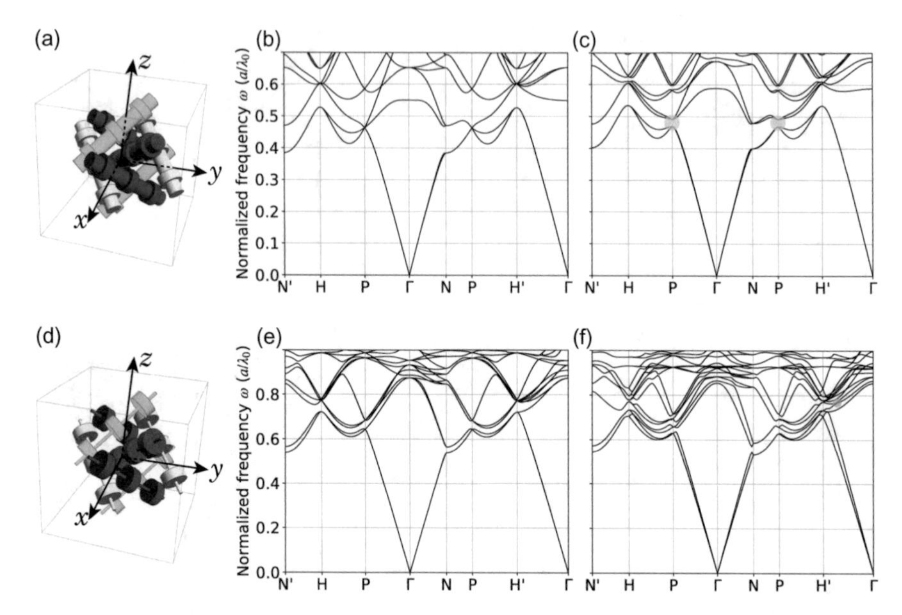


**Fig. 6.6** Configurations of photonic crystals with dielectric cylinders and the corresponding band structures with and without time-reversal symmetry (TRS). **a** When the dielectric cylinders at the Wyckoff position 32*e* are dominant, the band structures **b** with and **c** without TRS is similar to those of the BPI photonic crystal. **d** If the dielectric cylinders at the Wyckoff position 16*b* are dominant, the band structures **e** with and **f** without TRS generate those of the double gyroid photonic crystal

that of the double gyroid photonic crystal, and the Weyl nodes along $\Gamma$-$H'$ are the only band touching points between the fourth and fifth bands in the absence of TRS. To break TRS, we set $\varepsilon_{zz} = 12$ and $\kappa = 11i$ in Eq. (6.1) for all dielectric spheres.

Instead of dielectric spheres which are generally difficult to realize, we also consider dielectric cylinders. To generate the Wyckoff position 32*e* from 16*a* and 16*b*, we keep its site symmetry 3. Namely, we fix the directions of dielectric cylinders going along the axes connecting the sites 16*a* and 16*b*, which are corresponding to the BPI photonic crystal. Since the aim here is to show the interplay of the band structure between 32*e* and 16*b*, we combine two species of cylinders, one at 32*e* and the other at 16*b*. We fix the height and radius of dielectric cylinders at 16*b* as $h_b = \sqrt{3}/12$ and $r_b = 0.15$, and change the radius $r_e$ of dielectric cylinders at 32*e* with an infinite height (Fig. 6.6a, d). In the case of $r_e = 0.1$ which is almost the BPI photonic crystal shown in Fig. 6.6a, the band structure with TRS in Fig. 6.6b is similar to that of the BPI photonic crystal, as we intuitively expected. In Fig. 6.6c, the magnetization is present where $\kappa = -11i, -11i, 11i, 11i$ for the red, yellow, blue, and green cylinders, respectively, and the system goes into the glide-$Z_2$ TCI phase by opening the gap at the $P$ point. In the case of $r_e = 0.02$ where the Wyckoff position $16b$ is dominant in Fig. 6.6d, the band structure with TRS is shown in Fig. 6.6e which has double-gyroid-like band structure. By magnetization with $\kappa = 11i$ for all the cylinders, the fourth and fifth bands only cross each other at Weyl nodes in Fig. 6.6f.

## 6.4 Conclusion and Discussion

We study the relationship between band structures and space group representations to manipulate topological photonic crystals ensured by glide symmetry. We have shown that the BPI photonic crystal without time-reversal symmetry is the glide-$Z_2$ magnetic topological crystalline insulator phase by using our new formula of the glide-$Z_2$ invariant calculated from irreducible representations. In this calculation, one has to pay attention to the $\Gamma$ point due to its singularity in three-dimensional photonic crystals. We have also figured out the condition that the photonic crystal with ***230*** realizes the glide-$Z_2$ magnetic topological phase only from symmetry considerations. We found that dielectrics located at Wyckoff position $32e$ of ***230*** are a key to make the system topologically nontrivial. Our numerical calculations verify our scenario.

There remain several issues in this chapter. First, the gap induced by breaking time-reversal symmetry is very small in general. Besides, due to the threefold rotation along $\Gamma$-$P$-$H$ in this space group, the lowest frequency at the $H$ point is always higher than that at the $P$ point when we treat the photonic crystal perturbatively to the air. Namely, there is no photonic band gap in general even though the gap opens at the $P$ point, because the upper mode above the gap at the $P$ point has generally lower frequency than the lower mode below the gap at the $H$ point. Nevertheless, the condition for the glide-$Z_2$ topological photonic crystal that we propose is straightforward; the lowest bands with the irreducible representation $H_1$ at the $H$ point. Then, by opening the gap between the second and third bands by any means with breaking time-reversal symmetry, we always obtain the glide-$Z_2$ topological crystalline insulator.

Second, due to the singularity at the $\Gamma$ point, we cannot apply the theories for symmetry-based indicators or Wyckoff positions related to atomic insulators to photonic crystals in general as mentioned in Ref. [13]. As we pointed out earlier, a given crystal structure consistent with the corresponding space group is classified by Wyckoff positions, and by assuming tight-binding orbitals we generate a set of irreducible representations for an atomic insulator. However, besides the existence of the singularity at the $\Gamma$ point, the electromagnetic waves in photonic crystals propagate in vacuum, which means that one cannot describe photonic crystals in terms of tight-binding models. Thus, it is a future work to manifest one-to-one correspondence between the site symmetry for Wyckoff positions and the band structures and topological properties in photonic crystals.

**Table 6.4** Summary of the reducible representations and decomposed irreducible representations at the 8 lowest bands in vacuum with a frequency $\Omega_0$ where $b = 2\pi/a$ with the lattice constant $a$ in ***230***

| $R$ | $\Omega_0$ | $E$ | $3C_2$ | $8C_3$ | $6C_4$ | $6C_2'$ | $I$ | $3IC_2$ | $8IC_3$ | $6IC_4$ | $6IC_2'$ | Irreducible representations |
|---|---|---|---|---|---|---|---|---|---|---|---|---|
| $\chi_{\mathcal{R}(\Gamma^{(1)})}(R)$ | $\sqrt{2}b$ | 24 | 0 | 0 | 0 | −24 | 0 | 0 | 0 | 0 | 0 | $\Gamma_2^+ + \Gamma_2^- + \Gamma_3^+ + \Gamma_3^- + 2\Gamma_4^+ + 2\Gamma_4^- + \Gamma_5^+ + \Gamma_5^-$ |
| $\chi_{\mathcal{R}(H^{(0)})}(R)$ | $b$ | 12 | −12 | 0 | 0 | 0 | 0 | 0 | 0 | 0 | 0 | $H_1 + H_2 + H_3 + H_4$ |
| $\chi_{\mathcal{R}(H^{(1)})}(R)$ | $\sqrt{3}b$ | 16 | 0 | −16 | 0 | 0 | 0 | 0 | 0 | 0 | 0 | $H_2 + H_3 + 2H_4$ |
| $\chi_{\mathcal{R}(N^{(0)})}(R)$ | $\frac{\sqrt{2}}{2}b$ | 4 | 0 | – | – | 4 | 0 | 0 | – | – | 0 | $2N_1$ |
| $\chi_{\mathcal{R}(N^{(1)})}(R)$ | $\frac{\sqrt{6}}{2}b$ | 8 | 0 | – | – | 0 | 0 | 0 | – | – | 0 | $2N_1 + 2N_2$ |
| $\chi_{\mathcal{R}(N^{(2)})}(R)$ | $\frac{\sqrt{10}}{2}b$ | 8 | 0 | – | – | 0 | 0 | 0 | – | – | 0 | $2N_1 + 2N_2$ |
| $\chi_{\mathcal{R}(P^{(0)})}(R)$ | $\frac{\sqrt{3}}{2}b$ | 8 | 0 | −8 | – | – | – | – | – | 0 | 0 | $P_1 + P_2 + P_3$ |
| $\chi_{\mathcal{R}(P^{(1)})}(R)$ | $\frac{\sqrt{11}}{2}b$ | 24 | 0 | 0 | – | – | – | – | – | 0 | 0 | $2P_1 + 2P_2 + 4P_3$ |

**Table 6.5** Compatibility relations with time-reversal symmetry between the $P$ point and the $H$ point via $\Lambda$ which is along the threefold-rotation invariant axis. The full information is summarized in Ref. [12]

| Maximal $k$-vector | Compatibility relations | Intermediate path | Compatibility relations | Maximal $k$-vector |
|---|---|---|---|---|
| $P$ | $P_1P_2 \to 2\Lambda_3$<br>$P_3 \to \Lambda_1 \oplus \Lambda_2 \oplus \Lambda_3$ | $\Lambda$ | $H_1 \to \Lambda_1 \oplus \Lambda_2$<br>$H_2H_3 \to 2\Lambda_3$<br>$H_4 \to \Lambda_1 \oplus \Lambda_2 \oplus 2\Lambda_3$ | $H$ |

Our theory may also applied to magnons in yttrium iron garnet (YIG), a magnetic material belonging to the same space group ***230***. According to [16], the degree of degeneracy for the lowest bands at the $H$ point in the magnon band structure looks higher than four and the behavior of the band structures is different from photonic crystals we have discussed so far. Moreover, it is a challenging problem to control magnon spectrum, for example to open the gap or to invert the modes. Therefore, it remains an open question how our theory applies other quasi-particle systems.

## Supplementary Materials

Here we summarize useful information for understanding the main text.

Table 6.4 shows the results of Sect. 6.2.1. For the 8 lowest bands in vacuum with a frequency $\Omega_0$ in ***230***, we calculate the reducible representations and decomposed irreducible representations at the high-symmetry points.

We show characters of irreducible representations at high-symmetry points $H$ and $N$ in Table 6.6 and the compatibility relations with time-reversal symmetry between the $P$ point and the $H$ point via $\Lambda$ in Table 6.5 in ***230*** [12]. The full information is summarized in the database in Ref. [12].

The Wyckoff positions in ***230*** are shown in Table 6.7 [6]. The multiplicity and the Wyckoff letter are shown in the first column, the site symmetry is shown in the second column, and the coordinates with two sets $(0, 0, 0)+$ and $(\frac{1}{2}, \frac{1}{2}, \frac{1}{2})+$ are given in the rest of Table 6.7. The details are given in Ref. [6].

**Table 6.6** Summary of characters of irreducible representations at high-symmetry points $H$ and $N$ adapted from Ref. [12]

| | $H_1$ | $H_2$ | $H_3$ | $H_4$ | $N_1$ | $N_2$ | | $H_1$ | $H_2$ | $H_3$ | $H_4$ | $N_1$ | $N_2$ |
|---|---|---|---|---|---|---|---|---|---|---|---|---|---|
| $\{1\|t_1t_2t_3\}$ | 2 | 2 | 2 | 6 | 2 | 2 | $\{\bar{1}\|000\}$ | 0 | 0 | 0 | 0 | 0 | 0 |
| $\{2_{001}\|0\frac{1}{2}0\}$ | −2 | −2 | −2 | 2 | 0 | 0 | $\{m_{001}\|0\frac{1}{2}0\}$ | 0 | 0 | 0 | 0 | 0 | 0 |
| $\{2_{010}\|\frac{1}{2}00\}$ | −2 | −2 | −2 | 2 | | | $\{m_{010}\|\frac{1}{2}00\}$ | 0 | 0 | 0 | 0 | | |
| $\{2_{100}\|00\frac{1}{2}\}$ | −2 | −2 | −2 | 2 | | | $\{m_{100}\|00\frac{1}{2}\}$ | 0 | 0 | 0 | 0 | | |
| $\{3^+_{111}\|000\}$ | 2 | −1 | −1 | 0 | | | $\{\bar{3}^+_{111}\|000\}$ | 0 | $-\sqrt{3}i$ | $\sqrt{3}i$ | 0 | | |
| $\{3^+_{\bar{1}1\bar{1}}\|00\frac{1}{2}\}$ | −2 | 1 | 1 | 0 | | | $\{\bar{3}^+_{\bar{1}1\bar{1}}\|00\frac{1}{2}\}$ | 0 | $\sqrt{3}i$ | $-\sqrt{3}i$ | 0 | | |
| $\{3^+_{1\bar{1}\bar{1}}\|0\frac{1}{2}0\}$ | −2 | 1 | 1 | 0 | | | $\{\bar{3}^+_{1\bar{1}\bar{1}}\|0\frac{1}{2}0\}$ | 0 | $\sqrt{3}i$ | $-\sqrt{3}i$ | 0 | | |
| $\{3^+_{\bar{1}\bar{1}1}\|\frac{1}{2}00\}$ | −2 | 1 | 1 | 0 | | | $\{\bar{3}^+_{\bar{1}\bar{1}1}\|\frac{1}{2}00\}$ | 0 | $\sqrt{3}i$ | $-\sqrt{3}i$ | 0 | | |
| $\{3^-_{111}\|000\}$ | 2 | −1 | −1 | 0 | | | $\{\bar{3}^-_{111}\|000\}$ | 0 | $\sqrt{3}i$ | $-\sqrt{3}i$ | 0 | | |
| $\{3^-_{1\bar{1}\bar{1}}\|\frac{1}{2}00\}$ | −2 | 1 | 1 | 0 | | | $\{\bar{3}^-_{1\bar{1}\bar{1}}\|\frac{1}{2}00\}$ | 0 | $-\sqrt{3}i$ | $\sqrt{3}i$ | 0 | | |
| $\{3^-_{\bar{1}\bar{1}1}\|00\frac{1}{2}\}$ | −2 | 1 | 1 | 0 | | | $\{\bar{3}^-_{\bar{1}\bar{1}1}\|00\frac{1}{2}\}$ | 0 | $-\sqrt{3}i$ | $\sqrt{3}i$ | 0 | | |
| $\{3^-_{\bar{1}1\bar{1}}\|0\frac{1}{2}0\}$ | −2 | 1 | 1 | 0 | | | $\{\bar{3}^-_{\bar{1}1\bar{1}}\|0\frac{1}{2}0\}$ | 0 | $-\sqrt{3}i$ | $\sqrt{3}i$ | 0 | | |
| $\{2_{110}\|\frac{3}{4}\frac{1}{4}\frac{1}{4}\}$ | 0 | 0 | 0 | 0 | 2 | −2 | $\{m_{110}\|\frac{3}{4}\frac{1}{4}\frac{1}{4}\}$ | 0 | 0 | 0 | 0 | 0 | 0 |
| $\{2_{1\bar{1}0}\|\frac{1}{4}\frac{1}{4}\frac{1}{4}\}$ | 0 | 0 | 0 | 0 | 0 | 0 | $\{m_{1\bar{1}0}\|\frac{1}{4}\frac{1}{4}\frac{1}{4}\}$ | 0 | 0 | 0 | 0 | 0 | 0 |
| $\{4^-_{001}\|\frac{1}{4}\frac{1}{4}\frac{3}{4}\}$ | 0 | 0 | 0 | 0 | | | $\{\bar{4}^-_{001}\|\frac{1}{4}\frac{1}{4}\frac{3}{4}\}$ | 0 | 0 | 0 | 0 | | |
| $\{4^+_{001}\|\frac{1}{4}\frac{3}{4}\frac{1}{4}\}$ | 0 | 0 | 0 | 0 | | | $\{\bar{4}^+_{001}\|\frac{1}{4}\frac{3}{4}\frac{1}{4}\}$ | 0 | 0 | 0 | 0 | | |
| $\{4^-_{100}\|\frac{3}{4}\frac{1}{4}\frac{1}{4}\}$ | 0 | 0 | 0 | 0 | | | $\{\bar{4}^-_{100}\|\frac{3}{4}\frac{1}{4}\frac{1}{4}\}$ | 0 | 0 | 0 | 0 | | |
| $\{2_{011}\|\frac{1}{4}\frac{3}{4}\frac{1}{4}\}$ | 0 | 0 | 0 | 0 | | | $\{m_{011}\|\frac{1}{4}\frac{3}{4}\frac{1}{4}\}$ | 0 | 0 | 0 | 0 | | |
| $\{2_{01\bar{1}}\|\frac{1}{4}\frac{1}{4}\frac{1}{4}\}$ | 0 | 0 | 0 | 0 | | | $\{m_{01\bar{1}}\|\frac{1}{4}\frac{1}{4}\frac{1}{4}\}$ | 0 | 0 | 0 | 0 | | |
| $\{4^+_{100}\|\frac{1}{4}\frac{1}{4}\frac{3}{4}\}$ | 0 | 0 | 0 | 0 | | | $\{\bar{4}^+_{100}\|\frac{1}{4}\frac{1}{4}\frac{3}{4}\}$ | 0 | 0 | 0 | 0 | | |
| $\{4^+_{010}\|\frac{3}{4}\frac{1}{4}\frac{1}{4}\}$ | 0 | 0 | 0 | 0 | | | $\{\bar{4}^+_{010}\|\frac{3}{4}\frac{1}{4}\frac{1}{4}\}$ | 0 | 0 | 0 | 0 | | |
| $\{2_{101}\|\frac{1}{4}\frac{1}{4}\frac{3}{4}\}$ | 0 | 0 | 0 | 0 | | | $\{m_{101}\|\frac{1}{4}\frac{1}{4}\frac{3}{4}\}$ | 0 | 0 | 0 | 0 | | |
| $\{4^-_{010}\|\frac{1}{4}\frac{3}{4}\frac{1}{4}\}$ | 0 | 0 | 0 | 0 | | | $\{\bar{4}^-_{010}\|\frac{1}{4}\frac{3}{4}\frac{1}{4}\}$ | 0 | 0 | 0 | 0 | | |
| $\{2_{\bar{1}01}\|\frac{1}{4}\frac{1}{4}\frac{1}{4}\}$ | 0 | 0 | 0 | 0 | | | $\{m_{\bar{1}01}\|\frac{1}{4}\frac{1}{4}\frac{1}{4}\}$ | 0 | 0 | 0 | 0 | | |

**Table 6.7** Summary of Wyckoff positions in ***230*** except for the most general one $96h$. The first column denotes the multiplicity and the Wyckoff letter, the second column denotes site symmetry, and the coordinates with two sets $(0, 0, 0)+$ and $(\frac{1}{2}, \frac{1}{2}, \frac{1}{2})+$. Asymmetric unit is given by $-\frac{1}{8} \le x \le \frac{1}{8}; -\frac{1}{8} \le y \le \frac{1}{8}; 0 \le z \le \frac{1}{4}; \max(x, -x, y, -y) \le z$ adapted from Ref. [6]

| $48g$ | $..2$ | $\frac{1}{8}, y, \bar{y}+\frac{1}{4}$ | | $\frac{3}{8}, \bar{y}, \bar{y}+\frac{3}{4}$ | | $\frac{7}{8}, y+\frac{1}{2}, y+\frac{1}{4}$ | | $\frac{5}{8}, \bar{y}+\frac{1}{2}, y+\frac{3}{4}$ | |
|---|---|---|---|---|---|---|---|---|---|
| | | $\bar{y}+\frac{1}{4}, \frac{1}{8}, y$ | | $\bar{y}+\frac{3}{4}, \frac{3}{8}, \bar{y}$ | | $y+\frac{1}{4}, \frac{7}{8}, y+\frac{1}{2}$ | | $y+\frac{3}{4}, \frac{5}{8}, \bar{y}+\frac{1}{2}$ | |
| | | $y, \bar{y}+\frac{1}{4}, \frac{1}{8}$ | | $\bar{y}, \bar{y}+\frac{3}{4}, \frac{3}{8}$ | | $y+\frac{1}{2}, y+\frac{1}{4}, \frac{7}{8}$ | | $\bar{y}+\frac{1}{2}, y+\frac{3}{4}, \frac{5}{8}$ | |
| | | $\frac{7}{8}, \bar{y}, y+\frac{3}{4}$ | | $\frac{5}{8}, y, y+\frac{1}{4}$ | | $\frac{1}{8}, \bar{y}+\frac{1}{2}, \bar{y}+\frac{3}{4}$ | | $\frac{3}{8}, y+\frac{1}{2}, \bar{y}+\frac{1}{4}$ | |
| | | $y+\frac{3}{4}, \frac{7}{8}, \bar{y}$ | | $y+\frac{1}{4}, \frac{5}{8}, y$ | | $\bar{y}+\frac{3}{4}, \frac{1}{8}, \bar{y}+\frac{1}{2}$ | | $\bar{y}+\frac{1}{4}, \frac{3}{8}, y+\frac{1}{2}$ | |
| | | $\bar{y}, y+\frac{3}{4}, \frac{7}{8}$ | | $y, y+\frac{1}{4}, \frac{5}{8}$ | | $\bar{y}+\frac{1}{2}, \bar{y}+\frac{3}{4}, \frac{1}{8}$ | | $y+\frac{1}{2}, \bar{y}+\frac{1}{4}, \frac{3}{8}$ | |
| $48f$ | $2..$ | $x, 0, \frac{1}{4}$ | | $\bar{x}+\frac{1}{2}, 0, \frac{3}{4}$ | | $\frac{1}{4}, x, 0$ | | $\frac{3}{4}, \bar{x}+\frac{1}{2}, 0$ | |
| | | $0, \frac{1}{4}, x$ | | $0, \frac{3}{4}, \bar{x}+\frac{1}{2}$ | | $\frac{3}{4}, x+\frac{1}{4}, 0$ | | $\frac{3}{4}, \bar{x}+\frac{3}{4}, \frac{1}{2}$ | |
| | | $x+\frac{3}{4}, \frac{1}{2}, \frac{1}{4}$ | | $\bar{x}+\frac{1}{4}, 0, \frac{1}{4}$ | | $0, \frac{1}{4}, \bar{x}+\frac{1}{4}$ | | $\frac{1}{2}, \frac{1}{4}, x+\frac{3}{4}$ | |
| | | $\bar{x}, 0, \frac{3}{4}$ | | $x+\frac{1}{2}, 0, \frac{1}{4}$ | | $\frac{3}{4}, \bar{x}, 0$ | | $\frac{1}{4}, x+\frac{1}{2}, 0$ | |
| | | $0, \frac{3}{4}, \bar{x}$ | | $0, \frac{1}{4}, x+\frac{1}{2}$ | | $\frac{1}{4}, \bar{x}+\frac{3}{4}, 0$ | | $\frac{1}{4}, x+\frac{1}{4}, \frac{1}{2}$ | |
| | | $\bar{x}+\frac{1}{4}, \frac{1}{2}, \frac{3}{4}$ | | $x+\frac{3}{4}, 0, \frac{3}{4}$ | | $0, \frac{3}{4}, x+\frac{3}{4}$ | | $\frac{1}{2}, \frac{3}{4}, \bar{x}+\frac{1}{4}$ | |
| $32e$ | $.3.$ | $x, x, x$ | | $\bar{x}+\frac{1}{2}, \bar{x}, x+\frac{1}{2}$ | | $\bar{x}, x+\frac{1}{2}, \bar{x}+\frac{1}{2}$ | | $x+\frac{1}{2}, \bar{x}+\frac{1}{2}, \bar{x}$ | |
| | | $x+\frac{3}{4}, x+\frac{1}{4}, \bar{x}+\frac{1}{4}$ | | $\bar{x}+\frac{3}{4}, \bar{x}+\frac{3}{4}, \bar{x}+\frac{3}{4}$ | | $x+\frac{1}{4}, \bar{x}+\frac{1}{4}, x+\frac{3}{4}$ | | $\bar{x}+\frac{1}{4}, x+\frac{3}{4}, x+\frac{1}{4}$ | |
| | | $\bar{x}, \bar{x}, \bar{x}$ | | $x+\frac{1}{2}, x, \bar{x}+\frac{1}{2}$ | | $x, \bar{x}+\frac{1}{2}, x+\frac{1}{2}$ | | $\bar{x}+\frac{1}{2}, x+\frac{1}{2}, x$ | |
| | | $\bar{x}+\frac{1}{4}, \bar{x}+\frac{3}{4}, x+\frac{3}{4}$ | | $x+\frac{1}{4}, x+\frac{1}{4}, x+\frac{1}{4}$ | | $\bar{x}+\frac{3}{4}, x+\frac{3}{4}, \bar{x}+\frac{1}{4}$ | | $x+\frac{3}{4}, \bar{x}+\frac{1}{4}, \bar{x}+\frac{3}{4}$ | |
| $24d$ | $\bar{4}..$ | $\frac{3}{8}, 0, \frac{1}{4}$ | $\frac{1}{8}, 0, \frac{3}{4}$ | $\frac{1}{4}, \frac{3}{8}, 0$ | $\frac{3}{4}, \frac{1}{8}, 0$ | $0, \frac{1}{4}, \frac{3}{8}$ | $0, \frac{3}{4}, \frac{1}{8}$ | | |
| | | $\frac{3}{4}, \frac{5}{8}, 0$ | $\frac{3}{4}, \frac{3}{8}, \frac{1}{2}$ | $\frac{1}{8}, \frac{1}{2}, \frac{1}{4}$ | $\frac{7}{8}, 0, \frac{1}{4}$ | $0, \frac{1}{4}, \frac{7}{8}$ | $\frac{1}{2}, \frac{1}{4}, \frac{1}{8}$ | | |
| $24c$ | $2.22$ | $\frac{1}{8}, 0, \frac{1}{4}$ | $\frac{3}{8}, 0, \frac{3}{4}$ | $\frac{1}{4}, \frac{1}{8}, 0$ | $\frac{3}{4}, \frac{3}{8}, 0$ | $0, \frac{1}{4}, \frac{1}{8}$ | $0, \frac{3}{4}, \frac{3}{8}$ | | |
| | | $\frac{7}{8}, 0, \frac{3}{4}$ | $\frac{5}{8}, 0, \frac{1}{4}$ | $\frac{3}{4}, \frac{7}{8}, 0$ | $\frac{1}{4}, \frac{5}{8}, 0$ | $0, \frac{3}{4}, \frac{7}{8}$ | $0, \frac{1}{4}, \frac{5}{8}$ | | |
| $16b$ | $.32$ | $\frac{1}{8}, \frac{1}{8}, \frac{1}{8}$ | $\frac{3}{8}, \frac{7}{8}, \frac{5}{8}$ | $\frac{7}{8}, \frac{5}{8}, \frac{3}{8}$ | $\frac{5}{8}, \frac{3}{8}, \frac{7}{8}$ | $\frac{7}{8}, \frac{7}{8}, \frac{7}{8}$ | $\frac{5}{8}, \frac{1}{8}, \frac{3}{8}$ | $\frac{1}{8}, \frac{3}{8}, \frac{5}{8}$ | $\frac{3}{8}, \frac{5}{8}, \frac{1}{8}$ |
| $16a$ | $.\bar{3}.$ | $0, 0, 0$ | $\frac{1}{2}, 0, \frac{1}{2}$ | $0, \frac{1}{2}, \frac{1}{2}$ | $\frac{1}{2}, \frac{1}{2}, 0$ | $\frac{3}{4}, \frac{1}{4}, \frac{1}{4}$ | $\frac{3}{4}, \frac{3}{4}, \frac{3}{4}$ | $\frac{1}{4}, \frac{1}{4}, \frac{3}{4}$ | $\frac{1}{4}, \frac{3}{4}, \frac{1}{4}$ |

## References

1. Fang C, Fu L (2015) Phys Rev B 91:161105(R). https://doi.org/10.1103/PhysRevB.91.161105
2. Shiozaki K, Sato M, Gomi K (2015) Phys Rev B 91:155120. https://doi.org/10.1103/PhysRevB.91.155120
3. Shiozaki K, Sato M, Gomi K (2018) arXiv preprint arXiv:1802.06694
4. Lu L, Fang C, Fu L, Johnson SG, Joannopoulos JD, Soljačić M (2016) Nat Phys 12(4):337
5. Lu L, Fu L, Joannopoulos JD, Soljačić M (2013) Nat. Photonics 7(4):294
6. Hahn T (2002) International tables for crystallography, volume a: space group symmetry, 5th edn. International tables for crystallography. Kluwer Academic Publishers, Dordrecht, Boston, London
7. Haldane FDM, Raghu S (2008) Phys Rev Lett 100:013904. https://doi.org/10.1103/PhysRevLett.100.013904
8. Wang Z, Chong Y, Joannopoulos JD, Soljačić M (2009) Nature 461(7265):772
9. Kane CL, Mele EJ (2005) Phys Rev Lett 95:146802. https://doi.org/10.1103/PhysRevLett.95.146802
10. Kane CL, Mele EJ (2005) Phys Rev Lett 95:226801. https://doi.org/10.1103/PhysRevLett.95.226801
11. Fu L, Kane CL (2007) Phys Rev B 76:045302. https://doi.org/10.1103/PhysRevB.76.045302
12. Aroyo MI, Perez-Mato JM, Capillas C, Kroumova E, Ivantchev S, Madariaga G, Kirov A, Wondratschek H (2006) Zeitschrift für Kristallographie-Crystalline Materials 221(1):15
13. Watanabe H, Lu L (2018) Phys Rev Lett 121:263903. https://doi.org/10.1103/PhysRevLett.121.263903
14. Wohlgemuth M, Yufa N, Hoffman J, Thomas EL (2001) Macromolecules 34(17):6083
15. Joannopoulos JD, Johnson SG, Winn JN, Meade RD (2008) Molding the flow of light
16. Princep AJ, Ewings RA, Ward S, Tóth S, Dubs C, Prabhakaran D, Boothroyd AT (2017) npj Quantum Mater 2(1):63

# Chapter 7
# Conclusion and Outlook

In this thesis, we have studied $Z_2$ magnetic topological crystalline insulators ensured by glide symmetry. We established a general theory of the topological phase transition, new formulas of the glide-$Z_2$ topological invariant in the presence of inversion symmetry from both approaches in $k$-space and real-space, and a manipulation for such glide-symmetric $Z_2$ magnetic topological phase.

In Chap. 3, we presented that the spinless Weyl semimetal phase should emerge between $Z_2$ magnetic topological crystalline insulator and normal insulator phases in glide-symmetric systems. We first construct a general theory of the phase transition between the glide-$Z_2$ magnetic topological crystalline insulator and the normal insulator phases and we show a generic phase diagram involving the spinless Weyl semimetal phase using the effective model. In this case, the trajectory of the Weyl nodes within the Weyl semimetal phase governs the change of the glide-$Z_2$ topological invariant. In particular, when the Weyl nodes are created and annihilated pairwise in the sectors with opposite signs of glide eigenvalues, the glide-$Z_2$ topological invariant changes between the two sides of the Weyl semimetal phase. These scenarios are confirmed by our spinless tight-binding model with glide symmetry. In this model, we also show that surface Fermi arcs in the spinless Weyl semimetal phase evolve into a single surface Dirac cone in the glide-$Z_2$ magnetic topological crystalline insulator phase. We believe our results will be useful for materials search of the glide-$Z_2$ magnetic topological crystalline insulators and the spinless Weyl semimetals since it does not suffer from the constraint of significant spin-orbit coupling unlike $Z_2$ topological insulators. Magnetic insulators, such as insulators in magnetic field of localized spin systems, will be promising candidates. Magnonic systems are also good candidates due to the absence of time-reversal symmetry and due to the similarity on statistics between spinless and bosonic systems.

In Chap. 4, we studied the fate of the glide-$Z_2$ topological invariant in the presence of inversion symmetry additionally while time-reversal symmetry is not enforced. First, there is a need to redefine the glide-$Z_2$ topological invariant proposed in previous works to agree with the glide invariant constructed by layer constructions in

 
H. Kim, *Glide-Symmetric $Z_2$ Magnetic Topological Crystalline Insulators*,
Springer Theses, https://doi.org/10.1007/978-981-16-9077-8_7

real space in the following chapter. By redefinition, we showed that the glide-$Z_2$ topological invariant in the space groups No. 13 and No. 15 is solely expressed in terms of the irreducible representations at high-symmetry points in momentum space. It constitutes the symmetry-based indicators for these space groups, together with the Chern number modulo 2. In the space group No. 14, the combination of the glide-$Z_2$ topological invariant and a half of the Chern number, corresponds to the symmetry-based indicator in this space group, and is calculated from the irreducible representations at high-symmetry points. Therefore, one has to calculate integrals of the Berry curvature in order to know each value of these topological numbers. Furthermore, the glide-$Z_2$ magnetic topological crystalline insulators are directly related to the higher-order topological insulators ensured by inversion symmetry from compatibility relations in these space groups.

In Chap. 5, we established a layer construction in magnetic systems where each layer is decorated by a two-dimensional Chern insulator as an independent approach. We constructed all invariants with respect to each space-group operation based solely on real-space geometric properties of the layers in a certain space group. As a result, the set of invariants for layer constructions exactly agrees with topological invariants defined in momentum space topology which we discussed in the previous chapter. From the set of elementary layer constructions, we also established corresponding tight-binding models in the space groups No. 13, No. 14, and No 15. The consequences of numerical calculations for the Berry phase and the surface band structures support our expectations by analytic calculations.

These consequences in Chaps. 4 and 5 will be promising to provoke studies on understanding the interplay between topological phases and symmetries, and mutual relationships between various topological phases. In addition, we believe that the knowledge of layer constructions in the magnetic systems will be useful for gaining insights for relating topological phases.

In Chap. 6, we focused on the relationship between space group representations and band structures to manipulate the glide-$Z_2$ magnetic topological crystalline insulator phases, and discussed its result in photonic crystals. By using our new formula, we calculate the glide-$Z_2$ topological invariant for the photonic crystal in the space group No. 230, and we manifested that the irreducible representation at $H$ point, one of the high-symmetry points in the body-centered cubic lattice, is the key to realize the glide-$Z_2$ magnetic topological phase. From our analysis, among Wyckoff positions in the space group No. 230, only one specific Wyckoff position is related to the glide-$Z_2$ topological phase while some others represent Weyl nodes in the absence of time-reversal symmetry. Based on our scenario, we manipulated the topological band structures by giving several configurations constructed from the Wyckoff positions. We believe that our results give a guideline to manipulate the glide-$Z_2$ magnetic topological crystalline insulators, even though there remain several tasks. Our idea on a manipulation of topological phases by using Wyckoff positions and site symmetry will be promising not only in electronic systems but also in bosonic systems as we have shown.

Because the glide symmetry is one of the fundamental symmetries in crystals, and it is contained in many space groups, we expect our general results are widely available for studies on topological crystalline materials. From the results in this thesis, it is also promising to discover new materials for topological crystalline insulators and higher-order topological insulators and to explore their versatile phenomena.

# Curriculum Vitae

## Education

- Ph.D., Department of Physics, Tokyo Institute of Technology, 4/17-3/20
- M.A., Department of Physics, Tokyo Institute of Technology, 4/15-3/17
- B.A., Department of Physics, Tokyo Institute of Technology, 4/11-3/15

## Research Experience

### Ph.D.

Advisor: Prof. Shuichi Murakami, 4/15 – 3/20

- Theoretical study of magnetic topological crystalline insulators


H. Kim, *Glide-Symmetric $Z_2$ Magnetic Topological Crystalline Insulators*,
Springer Theses, https://doi.org/10.1007/978-981-16-9077-8

Printed in the United States
by Baker & Taylor Publisher Services